RNAi

RNAi

Martin Latterich (Editor)

Faculty of Pharmacy
University of Montreal
Quebec, Canada

Taylor & Francis
Taylor & Francis Group

Published by:

Taylor & Francis Group

In US: 270 Madison Avenue
         New York, N Y 10016
In UK: 2 Park Square, Milton Park
         Abingdon, OX14 4RN

Library of Congress Cataloging-in-Publication Data

RNAi / Martin Latterich, editor.
        p.   ;   cm.
     Includes bibliographical references and index.
     ISBN 978–0–415–40950–6 (alk. paper)
   1. Small interfering RNA.   2. Gene silencing.   I. Latterich, Martin.
II. Title: RNA interference.
     [DNLM: 1. RNA Interference. QU 475 R627 2008]

QP623.5.S63R65 2008
572.8′8—dc22
                                              2007020070

Editor: Elizabeth Owen
Editorial Assistant: Kirsty Lyons
Production Editor: Karin Henderson
Typeset by: Phoenix Photosetting, Chatham, Kent, UK
Printed by: Cromwell Press, Trowbridge, Wiltshire, UK

Printed on acid-free paper

10 9 8 7 6 5 4 3 2 1

Taylor & Francis Group, an informa business

Visit our web site at http://www.garlandscience.com

Contents

Contributors

Leena Alhonen, Department of Biotechnology and Molecular Medicine, A.I. Virtanen Institute, University of Kuopio, PO Box 627, FIN-70211 Kuopio, Finland

Nathalie Aulner, Department of Physiology and Cellular Biophysics, Columbia University, 1150 St Nicholas Avenue, RB528, New York, NY 10032, USA

Eric Chevet, Team AVENIR, INSERM U889, IFR66, Université Bordeaux 2, 146 rue Léo Saignat, 33076 Bordeaux, France

Douglas S. Conklin, Gen*NY*Sis Center for Excellence in Cancer Genomics, University at Albany, State University of New York, Rensselaer, New York, NY 12144, USA

Cheryl Eifert, Gen*NY*Sis Center for Excellence in Cancer Genomics, University at Albany, State University of New York, Rensselaer, New York, NY 12144, USA

Dalia Halawani, Faculty of Pharmacy, University of Montreal, Montreal, Quebec, H3T 1J4, Canada

Jenni Huusko, Department of Biotechnology and Molecular Medicine, A.I. Virtanen Institute, University of Kuopio, PO Box 627, FIN-70211 Kuopio, Finland

Bernd Jagla, Department of Physiology and Cellular Biophysics, Columbia University, 1150 St Nicholas Avenue, RB528, New York, NY 10032, USA

Sarah Jenna, Team AVENIR, INSERM U889, IFR66, Université Bordeaux 2, 146 rue Léo Saignat, 33076 Bordeaux, France

Antonis Kourtidis, Gen*NY*Sis Center for Excellence in Cancer Genomics, University at Albany, State University of New York, Rensselaer, New York, NY 12144, USA

Martin Latterich, Faculty of Pharmacy, University of Montreal, Montreal, Quebec, H3T 1J4, Canada

Petri I. Mäkinen, Department of Biotechnology and Molecular Medicine, A.I. Virtanen Institute, University of Kuopio, PO Box 627, FIN-70211 Kuopio, Finland

Craig A. Mandato, Department of Anatomy and Cell Biology, McGill University, Montreal, QC, Canada

Hitoshi Nakayashiki, Laboratory of Plant Pathology, Kobe University, Kobe, Japan

Ouathek Ouerfelli, Organic Synthesis Core Laboratory, Molecular Pharmacology and Chemistry Program, The Sloan-Kettering Institute, Memorial Sloan-Kettering Cancer Center, 1275 York Avenue, New York, NY 10021, USA

Adrianna L. Stromme, Department of Anatomy and Cell Biology, McGill University, Montreal, QC, Canada

Seppo Ylä-Herttuala, Department of Biotechnology and Molecular Medicine, A.I. Virtanen Institute, University of Kuopio, PO Box 627, FIN-70211 Kuopio, Finland

Abbreviations

BAC	bacterial artificial chromosome		LH	luteinizing hormone
bDNA	branched DNA		LTR	long terminal repeat
CCT2	T-complex protein 1, β-subunit		LV	lentivirus vector
			MALDI-TOF	matrix-assisted laser desorption ionization–time of flight
cDNA	complementary DNA			
COPAS	Complex Object Parametric Analyzer and Sorter		MBT	mid-blastula transition
			MCS	multiple cloning site
cPPT	central polypurine tract		miRNA	micro-RNA
Dicer	intracellular endonuclease complex		MMR	Marc's Modified Ringer
			MPP	membrane-permeant peptides
DLBCL	diffuse large B-cell lymphoma		mRNA	messenger RNA
dNTP	Deoxyribonucleotide triphosphate		MSCV	murine stem cell virus
			MSUD	meiotic silencing by unpaired DNA
DPC	days postcoitum			
dsRNA	double-stranded RNA		NFκB	nuclear factor κ B
DTT	dithiothreitol		nt	nucleotide
ELISA	enzyme-linked immunosorbent assay		OAS1	2′,5′-oligoadenylate synthase
			ORF	open reading frame
ER	endoplasmic reticulum		PCR	polymerase chain reaction
ES	embryonic stem		PEG	polyethylene glycol
EXT	extinction		PKR	protein kinase R
FACS	fluorescence-activated cell sorting		PMSG	pregnant mare's serum gonadotropin
FLU1	green fluorescence emission		pri-miRNA	primary miRNA
FSH	follicle-stimulating hormone		PTGS	post-transcriptional gene silencing
FYCO1	FYVE and coiled coil containing protein 1		PVP	polyvinylpyrrolidone
GFP	green fluorescent protein		QD	quantum dots
hCG	human chorionic gonadotropin		*qde*	quelling-deficient
			qRT-PCR	quantitative RT-PCR
HDAC4	histone deacetylase 4		RDA	rhodamine
HTA-TIP	histone acetyl transferase TIP60		RdRP	RNA-dependent RNA polymerase
IL-8	interleukin-8		RISC	RNA-induced silencing complex
IRF-3	interferon regulatory factor-3			
JAK-STAT	Janus kinase–signal transducers and activators of transcription		RISC*	activated RISC
			RNA	ribonucleic acid
			RNA pol	RNA polymerase
LATS2	large tumor suppressor homologue 2		RNAi	RNA interference
			RNase	ribonuclease
LC-MS	liquid chromatography-mass spectrometry		RPA	RNA protection assay
			RSK4	ribosomal S6 kinase 4

RT	reverse transcriptase	**TOF**	time of flight
SAHS	S-adenosyl-L-homocysteine hydrolase	**TOM**	triisopropylsiloxymethyl
		tsLT	temperature-sensitive allele of SV40 large T antigen
shRNA	short hairpin RNA		
siRNA	small interfering RNA	**UPR**	unfolded protein response
SV40	simian virus 40	**UTR**	untranslated regions
TBDMS	tertiary-butyldimethylsilyl	**UV**	ultraviolet
TERT	telomerase catalytic subunit	**VSV**	vesicular stomatitis virus
TGCT	testicular germ cell tumor	**WPRE**	woodchuck hepatitis virus post-transcriptional regulatory element
THN	trihydroxynaphthalene reductase		
TNF-α	tumor necrosis factor-α		

Preface

There is hardly any recent discovery that has attracted as much attention as the application of RNA interference (RNAi) technology. Its application enabled investigators from hypothesis-driven research, as well as from high-throughput screening backgrounds, to rapidly study gene product function by specifically ablating a gene product of interest in multicellular organisms. Although young as a technique, RNA interference already has evolved beyond a simple gene function evaluation tool, having won a clear position in functional screening and become the method of choice in drug target and pathway validation, the construction of gene ablated animal models, and the design of therapeutic agents based on RNAi.

While RNA interference clearly has become one of the most powerful techniques in molecular biology since the advent of PCR, it is far less robust than PCR, especially considering issues pertaining to design and implementation in different systems. It is therefore somewhat difficult to provide methods 'cookbook-style', because often these methods are out of date as soon as they have been written. Instead, the book places the emphasis on explaining the rational basis for RNA interference. Specifically, I focussed the first part of the book on some theoretical aspects of RNAi probe design, probe synthesis and vector design. The goal is to provide the reader with enough theory and links to ample resources available in the community to make informed decisions on which method is best under what circumstances. The second part of the book broadly includes established and emerging organismal gene ablation approaches, such as in certain fungi, *Xenopus laevis* and *Mus musculus*. Lastly, some chapters focus on performing RNAi-based functional screens, allowing the interrogation of gene functions in specific pathways, as well as on the construction of RNAi ablated mouse models. In my opinion, these two areas of research are especially well poised for future developments, given their central position and utility in biomedical research.

This book has been written for novices and intermediate-level scientists who are not familiar with the concept of RNA interference, yet are in a position to apply it to their area of interest. It is meant as a general introduction to applying principles of RNAi to specific research topics, as well as to experimental systems outside of *Drosophila melanogaster* and mammalian tissue-culture cell systems. The authors have included up-to-date references and current links to RNAi resources available on the web.

Lastly, I wanted to take this opportunity to thank my many colleagues who pioneered work in this exciting area for their generous and timely contributions. Without them, it would not have been possible to complete this book. Special thanks go to Doug Conklin for his fruitful discussions during the conceptual stages of this work.

Martin Latterich

Methods in RNA interference

Martin Latterich and Dalia Halawani

Since the advent of DNA sequencing and polymerase chain reaction (PCR), there has been rarely one emerging technology that has received as much attention as the use of RNA interference (RNAi). The reason for this enthusiasm is that the phenomenon of RNAi firstly enabled a simple and inexpensive way to rapidly ablate specific messenger RNA (mRNA) species by inducing their degradation via a cellular protein machinery collectively named the RNA-induced silencing complex or RISC (Ketting *et al.*, 2001). The phenomenon of RNAi was well known as a mechanism of inducing post-transcription gene silencing in plants and bacteria, where anti-sense RNA is being used to artificially silence the translation of proteins in select species. However, it is the discovery that small RNA polymers of 19–23 nucleotides can post-transcriptionally interfere with gene expression, either by inhibiting translation or inducing the degradation of complementary mRNA strands, that opened up applications in post-genomic research. For the first time it is now possible to synthesize small RNA species, as single-stranded, double-stranded or small hairpin structures, and introduce these molecules through common transfection methods into cells, where they serve to guide the RNA degradation machinery to the select target species. The RISC complex then systematically degrades the complementary mRNA, effectively resulting in the ablation of a specific mRNA species. Depending on the efficiency of ablation and the stability of the corresponding protein, the RNAi will ultimately result in the loss or reduction of the gene product. One of the major attractions of this technology is that it enables functional genetic analyses in eukaryotic systems that have been previously resilient to rapid genetic study, as well as gene ablation screens.

While the field of RNAi studies is relatively young, it has witnessed a rapidly expanding number of publications (*Figure 1.1*), with a remarkable 9000 publications in only 8 years. We predict that RNAi will not only remain an effective research tool to investigate gene function, but will soon be exploited in therapeutic applications to quench the expression of undesirable gene products. It is noteworthy that the recent Nobel Prize in Physiology or Medicine was awarded to Andrew Fire and Craig Mello for originally discovering and unraveling the molecular mechanism of double stranded RNAi (Fire *et al.*, 1998), an outcome quite expected for a technique with such high impact.

The importance of RNAi and the wealth of published methods make it timely to compile a book with current techniques of use to the novice as well as expert user alike. We have decided to focus on some of the most

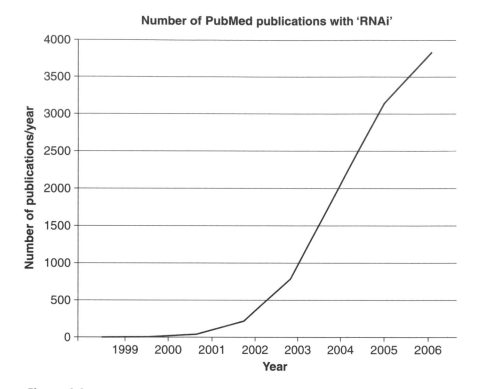

Figure 1.1

Publication trend in research focusing on RNAi. A PubMed search, using 'RNAi' as keyword and restricted by year of publication was performed, and number of publications containing 'RNAi' were plotted against the year of their publication.

common applications of RNAi, such as design, synthesis, and their introduction into different cell types and organisms, as well as some high throughput screening applications of emerging importance. While this book attempts to accurately convey current techniques, the methodologies are still evolving and innovation is much under way. We encourage the scientist user to handle the published methods as a basis for further experimentation and development. One key fact to remember is that while RNAi is a powerful tool, it has its limitations in terms of specificity and sensitivity. It is therefore necessary to validate the specificity of gene ablation using orthogonal approaches before reaching any conclusions.

References

Fire, A., Xu, S., Montgomery, M.K., Kostas, S.A., Driver, S.E. and Mello, C.C. (1998) Potent and specific genetic interference by double-stranded RNA in *Caenorhabditis elegans*. *Nature* **391**: 806–811.

Ketting, R.F., Fischer, S.E., Bernstein, E., Sijen, T., Hannon, G.J. and Plasterk, R.H. (2001) Dicer functions in RNA interference and in synthesis of small RNA involved in developmental timing in *C. elegans*. *Genes Devel* **15**: 2654–2659.

RNAi reagent design

Bernd Jagla and Nathalie Aulner

2

2.1 Introduction

Over the last decade, RNAi has emerged as a key technology to study gene function and perform functional genomics studies. RNAi is a gene-specific knockdown technology in which the degradation of the target mRNA is guided by a homologous double-stranded RNA (dsRNA). Design considerations of RNAi reagents are the focus of this chapter.

RNAi is a well conserved process that is found throughout the eukaryotic kingdom. It is believed to be part of the cellular defense mechanism and, in the case of micro-RNAs (miRNAs), an integral part of post-transcriptional gene regulation (Ambros, 2001). Many studies have outlined the general mechanism: a long dsRNA is cut into small 21–23 base pair molecules termed small interfering RNAs (siRNA) by an intracellular endonuclease called Dicer (Ketting *et al.*, 2001) before being loaded into RISC and targeting the specific degradation of an homologous mRNA (Martinez *et al.*, 2002).

The discovery by Tuschl and colleagues that 21–23mer RNA duplexes can bypass mammalian cellular defense mechanisms against dsRNAs (Elbashir *et al.*, 2001a) allowed the technology to quickly expand to mammalian cell systems. There are two major ways of introducing siRNAs into mammalian cells: (i) direct transfection of siRNA duplexes (chemically or enzymatically produced) and (ii) introduction of a vector driving the expression of short hairpin RNA (shRNA) that are further processed into siRNAs by Dicer. The double-stranded siRNA consists of a 'guide strand' or 'guide' and a 'passenger strand' or 'passenger'. The guide strand binds to RISC forming activated RISC (RISC*) and the passenger strand is degraded.

To ensure the specificity and efficacy of the active siRNA molecules, it is pertinent to develop strategies for their rational design. Several sources of information are available that help finding rules and guidelines to successfully design siRNAs. Among these, studies of crystal structures of the RNAi machinery are one of the most revealing; they allow us to understand some of the properties that make a siRNA effective on a molecular level. Additionally, statistical analyses of active and non-active siRNA sequences are being used to infer siRNA design rules.

The risk of unspecific responses is an important consideration when designing RNAi reagents. Unspecific responses can be minimized by optimizing the sequence, experimental conditions, or chemically modifying the siRNA.

Algorithms based on considerations of RISC, sample sequences, and structural features of siRNA have been developed and some of them are available through web services and/or standalone programs. In this chapter, we describe what can be learned from recent structural studies of important players in the RNAi process, current design considerations, tools and databases available on the World Wide Web, and other related resources that help designing RNAi probes with a high probability of being active.

2.2 Lessons learned from X-ray structures/mechanism

The RISC is composed of Argonaute 2 and the single-strand guide. It is responsible for recognizing and processing dsRNAs into siRNAs, recognizing the target mRNA and processing it (Filipowicz, 2005).

Crystal structures of RNA silencing complexes provide insights into the recognition and cleavage machinery. From these structures we know that (i) the recognition of siRNAs requires a phosphorylated 5'-terminus of the guide strand; (ii) the 5'-terminal nucleotide is not bound to the complementary strand but rather interacts with RISC; (iii) the cleavage of the passenger and target RNA occurs at the position localized between nucleotides 10 and 11 on the guide strand; and (iv) the nucleotides of the 3'-end are more loosely connected to RISC and are most probably not important for target recognition (Ma *et al.*, 2005; Parker *et al.*, 2005).

As a practical consequence, it is desirable to have unique nucleotides between positions 2 and 11 of the guide, thus avoiding silencing unintended targets or introducing 'off-target' effects, and having an A, U (lower energy base pair) or unpaired nucleotide at the 5'-end of the guide strand.

Insights into RISC-mediated mRNA cleavage were obtained by analyzing the crystal structure of *Archaeoglobus fulgidus* Piwi protein (*Af*-Piwi) complexed with dsRNA (Ma *et al.*, 2005; Parker *et al.*, 2004) and *Aquifex aeolicus* Argonaute (*Aa* Ago) (Yuan *et al.*, 2005). The *Aa* Ago protein is composed of four domains, an *N*-terminal domain, followed by a PAZ domain, a Mid domain, and a Piwi domain. The Piwi domain contains the mRNA cleavage site. The Mid domain contains the binding pocket for the 5' phosphate of the guide strand, and the PAZ domain contains a 3' overhang binding pocket. The process of guide strand-mediated mRNA binding, cleavage, and release within the context of the Ago scaffold in RISC comprises of a four-step catalytic cycle: nucleation, propagation, cleavage, and release (Yuan *et al.*, 2005). The 5'-end of the guide strand is anchored within a highly conserved pocket formed by the Mid domain and the Piwi domain, making this nucleotide inaccessible for base pairing. This explains why nucleotides with low binding energies (A/U) are preferred at this position (Jagla *et al.*, 2005). It also implies that this pocket is likely to be the site of 5'-end recognition of the guide strand and thus initiates nucleation. The Watson–Crick edges of nucleotides 2–8 (seed region) of the guide are directed outward into the solvent, providing access for target mRNA identification. Propagating the zippering up of the guide strand with target mRNA leads to the full-length duplex. Site-specific cleavage occurs between positions 10 and 11, as defined from the 5'-end of the guide strand, and uses a two-metal ion mechanism, like most type III ribonucleases (Parker *et al.*, 2004). These structural data are highly correlated to experimental data; it has been proposed that base pairing between the 5'-end of the guide RNA and the mRNA is more critical than pairing at the 3'-end (Doench and Sharp, 2004). In addition, Haley and Zamore have shown that up to nine contiguous non-canonical pairs can be tolerated within the duplex segment towards the 3'-end of the guide strand (Haley and Zamore, 2004). This can be explained by crystal structure analysis showing that this region is not required for the formation of stable crystals (Yuan *et al.*, 2005). Nonetheless, it appears that proper base pairing contributes to the catalytic rate (Haley and Zamore, 2004). Release of the cleaved product closes the cycle and frees RISC up to silence the next mRNA.

2.3 Current design considerations

Apart from the considerations that can be derived from the crystal structures, statistical analysis of experimentally verified functional and non-functional siRNAs can provide insights into the design of effective siRNAs. Important considerations for designing RNAi reagents are: asymmetry, positional preferences of certain nucleotides within the sequence, non-sequence position-based considerations, practical considerations relating to synthesis or experimental setup, and specificity. In the following section, we describe these design considerations in more detail. How to improve siRNAs by chemical modification is described elsewhere in this book (Chapter 3).

2.3.1 Asymmetry

A dsRNA is asymmetric if one side of the molecule has a higher binding energy than the other, that is, contains more G/C. When designing RNAi reagents, the 5'-end of the guide strand should have a lower binding energy than the 3'-end (Schwarz *et al.*, 2003). More specifically, the nucleotide at the 5'-end of the guide strand should be A, U or even could be unpaired with the target or the passenger strand. This will favor one strand to assemble into RISC and trigger the interference. It was shown that by changing the symmetry of an inactive siRNA, efficacy was increased (Hutvagner, 2005; Khvorova *et al.*, 2003; Schwarz *et al.*, 2003). Moreover, statistical analysis has shown that functional siRNA duplexes have a lower internal stability at the 5'-end of the guide than that of non-functional duplexes (Khvorova *et al.*, 2003; Reynolds *et al.*, 2004). Another way to ensure this asymmetry is to apply a combination of rules, based on statistical analysis. For example, work from our lab and those from Ui-Tei and colleagues have shown that at least five A or U residues in the 5'-terminal third of the guide strand are required (Jagla *et al.*, 2005; Ui-Tei *et al.*, 2004). In conjunction with a medium-high G/C content, this means that the 5'-end of the guide strand needs to have a higher A/C content than the 3'-end. Other flavors are described by Santoyo *et al.* (2005), who require between two and four A/U within the 3'-end of the passenger strand and a minimum number of G/C at the 5'-end of the same strand. Interestingly, it has been shown recently that there was a difference between mammalian and *Drosophila* systems; in the latter case, as you can introduce longer RNA duplexes which are cut in 21–23mers intracellularly by Dicer – there is no need for strand bias (Preall *et al.*, 2006).

2.3.2 Sequence positional preferences

Statistical analysis has shown that certain positions within the siRNA duplex contribute differently to the efficacy. Almost all of these analyses show that positions 1, 19, and 10 of the passenger strand have the most prominent effects. G/C at position 1, A/U at position 19, and A/U at position 10 are most favorable (Amarzguioui et al., 2005; Jagla et al., 2005; SVM RNAi; Chang Bioscience and Castro Valley; Ui-Tei et al., 2004). This can also be interpreted as an asymmetry feature and reasoned that since the nucleotide at the 5′-end of the guide strand has to be single stranded for incorporation into RISC (Yuan et al., 2005) a weaker bond would be preferred.

The cleavage site for the mRNA/passenger strand is located between positions 10 and 11 of the guide strand (Yuan et al., 2005). It is therefore not surprising that statistical analyses have shown certain preferences for this position. We and others showed that an A/U at position 10 is favorable (Jagla et al., 2005; Yoshinari et al., 2004). A, G, or C at position 11 seems to increase the chance of designing a functional siRNA (Hsieh et al., 2004).

Other positions, including position 3, where G/C is favorable (Amarzguioui and Prydz, 2004; Santoyo et al., 2005; SVM RNAi; Chang Bioscience and Castro Valley), position 16 (C) (Amarzguioui and Prydz, 2004), position 6 (A) (Amarzguioui and Prydz, 2004), and position 13 (U) (Amarzguioui and Prydz, 2004) have been shown to have a positive effect based on statistical analyses.

2.3.3 Practical features

Practical features include considerations to simplify the chemical synthesis of siRNAs or to disfavor potential internal loops. A long stretch of one nucleotide is a commonly excluded motif. This is based on the assumption that long stretches of the same nucleotide can pose problems either for synthesis or because of the potential formation of aggregates. The most prominent negative feature, which has been observed, is a stretch of four or more G in a row (Santoyo et al., 2005; Ui-Tei et al., 2004), but other nucleotides are also being considered. A more general way of formulating this rule is requiring a variation of 20% in nucleotide composition (Santoyo et al., 2005). The overall G/C content falls into the same category, as there is no biophysical proof that the overall G/C content should be in a certain range. Nonetheless, most statistical analyses propose maintaining a G/C content somewhere between 20% and 55% (Reynolds et al., 2004; Santoyo et al., 2005; SVM RNAi; Chang Bioscience and Castro Valley). Internal loops are unfavorable because of potential competing binding during duplex formation.

2.3.4 Non-sequence position-based considerations

Non-sequence position-based design considerations relate to the selection of an siRNA well within the coding region of the target mRNA, avoiding siRNAs that can form secondary structures, and ensuring specificity.

Usually, the search region for siRNAs within the target mRNA should be limited to the region included between the start and stop codon position on the target mRNA (Santoyo *et al.*, 2005). It is also advisable to avoid 5′ and 3′ untranslated regions (UTRs) as well as the direct neighborhood of the start and stop codons, because these regions are known to be rich in regulatory motifs, and UTR-binding proteins. The reasoning behind these considerations is that RNA binding proteins or highly structured RNA regions could interfere with siRNA binding.

Another important consideration for the design of the siRNAs is to check for its specificity against all the mRNAs of the target organism, because it has been shown that off-target effects occur with as few as one or as many as 12 mismatches (Birmingham *et al.*, 2006). Thus, it is critical to use the smallest possible window size when performing BLAST searches. For this purpose, the NCBI provides a special BLAST (Altschul *et al.*, 1990) interface for small nearly exact matches where the standard parameters can be used (http://www.ncbi.nlm.nih.gov/blast/). Unfortunately, apart from the window size, BLAST searches do not consider non-Watson–Crick base pairing. The A G:U wobble between the guide strand and the target mRNA has a negative free energy and, thus, preserves partial complementarity. In contrast, a purine:purine mismatch between the guide strand and the target mRNA has an unfavorable free energy, lacks complementarity and would hinder RISC binding and therefore RNAi (Aronin, 2006). The position of these mismatches might also have a direct impact on the efficacy. A study of a small set of siRNAs and targets has shown that mismatched nucleotide pairs at the ends of an siRNA conserve RNAi activity and are better tolerated compared with central mismatches (Du *et al.*, 2005). One also has to consider that the database used for blasting is in a constant flux and will always only reflect the latest version of the genome, and chances are that modifications will be made within the near future. Also, as described above, the binding energy between RISC* and the target mRNA is determining if an siRNA is active or not and a BLAST search can only be regarded as an approximation of this binding energy. Alternatives to BLAST searches would include optimized Smith–Waterman alignment algorithm (Birmingham *et al.*, 2006) or energy calculations based on Mfold (Mathews *et al.*, 1999b; Zuker, 1989) or similar programs. These approaches are computationally very expensive and it can be more practical sometimes to verify the siRNAs experimentally (see Chapter 4).

It is important to consider potential homologous sequences and gene families when designing an siRNA. Depending on the biological question asked, one should consider targeting all transcripts from a gene family or all transcripts coding for a protein with a particular functional domain. Blasting the target mRNA sequence will give homologous regions of the target and will give clues on where to look for potential off-targets. The specificity of the siRNA can also be improved by avoiding SNPs, exon boundaries, and repeat regions (Santoyo *et al.*, 2005). These regions are prone to sequencing errors, thus reducing the efficacy. Because of experimental constraints, it might be necessary to design an siRNA against one such region to target, for example, a specific SNP. In such cases, it is advisable to have the SNP positioned within the seed region (Parker *et al.*, 2005).

Finally, recent studies have suggested that the secondary structure of the target mRNA might have substantial influence on the siRNA efficacy (Schubert *et al.*, 2005). We will discuss the effects of secondary structures of the mRNA target and how to detect them later in this chapter.

Any siRNA sequence that is derived from a computer program or design rules should be considered as having an increased chance of being an active siRNA and should be confirmed experimentally. Databases collecting experimentally verified siRNAs are beginning to emerge. A comprehensive list of these databases is described later in this chapter.

2.3.5 shRNA design considerations

RNA interference can not only be induced by siRNA but also by shRNAs. shRNAs are encoded on plasmids or viruses and use the intracellular expression system and Dicer to generate siRNAs within the cell. Therefore the above-mentioned design rules can also be used to design shRNAs. In addition to siRNAs, however, shRNAs have a linker region and flanking promoter and termination regions (Miyagishi *et al.*, 2004). To drive the intracellular expression of the shRNA, RNA polymerase (pol) III and more recently RNA pol II promoters have been successfully used together with different termination sequences that can also be optimized to control the shRNA expression (Huppi *et al.*, 2005). Moreover, mammalian selection markers for long-term expression, the coexpression of marker transgenes from the same plasmid to ease identification of cells expressing the shRNA, and the inclusion of 'barcode DNA sequences' to identify siRNAs have been described (Amarzguioui *et al.*, 2005; Berns *et al.*, 2004). The targeted cell type and the length of time for which shRNA expression is required are among important factors to consider for choosing which type of expression system should be used to introduce the shRNA into the cells. Because cloning, bacterial amplification, and potentially virus preparation have the potential to induce sequence changes, which may influence both the efficacy and the specificity, the shRNA construct has to be confirmed by sequencing. This sequencing step can be difficult because of the intrinsic secondary structure of the hairpin coding sequence. One possible solution to overcome this shortcoming is to engineer restriction sites into the loop/stem region (Akeju *et al.*, 2006; Ducat *et al.*, 2003) or to modify the sequencing reaction (Taxman *et al.*, 2006). Other potential problems with shRNAs are the choice of promoter, orientation of guide and passenger strand, length of the RNA duplex (19–30 nucleotides; Kim *et al.*, 2005; Siolas *et al.*, 2005), loop structure, and the addition of a leader sequence (Huppi *et al.*, 2005; Paddison *et al.*, 2004b). Experimental studies have shown that: (i) RNA pol III promoters can be interchanged (Zeng *et al.*, 2002); (ii) 30-nucleotide hairpins are more effective than shorter sequences; (iii) the loop structure does not influence the efficacy; and (iv) that the addition of a U6 leader sequence has a positive effect (Paddison *et al.*, 2004b; Siolas *et al.*, 2005). Many strategies for designing shRNA loops and other specific aspects of shRNAs have been described over the past few years (Akeju *et al.*, 2006; Miyagishi *et al.*, 2004; Paddison *et al.*, 2004a, 2004b; Taxman *et al.*, 2006).

2.4 Prediction tools

Computational tools that predict potentially functional siRNAs only increase the probability of that siRNA being active. There are many programs available for the prediction of siRNAs, shRNAs, and miRNAs. They use different algorithms and try to solve different biological or technical problems related to the successful application of RNAi in different organisms. Here, we try to provide a comprehensive alphabetically sorted list of publicly available tools. As with all these 'comprehensive' lists, this list is only a snapshot of what is available at a given time point (May 2006).

- Ambion siRNA Target Finder (http://www.ambion.com/techlib/misc/siRNA_finder.html) implements the rules from Elbashir *et al.* (2001a, 2001b) and performs optional individual BLAST searches.
- BIOPREDsi (http://www.biopredsi.org/start.html) (Huesken *et al.*, 2005) uses an artificial neural network based on 2182 experimentally tested siRNAs and performs specificity tests for human, mouse, and rat.
- BLOCK-iT RNAi designer (http://rnaidesigner.invitrogen.com/sirna/) implements some proprietary rules. There are different tools available for designing miRNAs, siRNAs, and shRNAs. The program can blast sequence databases of different organisms to find unique regions on your sequence as potential targets.
- Clontech (http://bioinfo.clontech.com/rnaidesigner/sirnaSequenceDesign.do) implements some basic melting temperature calculations, uses some pattern rules, and performs optional individual BLAST searches.
- DEQOR (http://cluster-1.mpi-cbg.de/Deqor/deqor.html (Henschel *et al.*, 2004) applies symmetry analysis and G/C contents for selecting siRNAs. The program uses genomic sequences together with transcriptome analysis to check for cross-specificity. A command-line-based version of DEQOR is available.
- EMBOSS siRNA (http://athena.bioc.uvic.ca/cgi-bin/emboss.pl?_action=input&_app=siRNA) applies the rules from Elbashir *et al.* (2001a, 2001b), scores sequences according to G/C content, position in the mRNA sequence, and some sequence patterns. No specificity check is performed.
- E-RNAi web service (http://www.dkfz.de/signaling2/e-rnai/) (Arziman *et al.*, 2005) designs and evaluates dsRNA constructs suitable for RNAi experiments in invertebrate, *Drosophila* and *C. elegans*. dsRNA sequences (RNAi probes) are evaluated for their predicted specificity and efficiency. Since DNA templates used to generate dsRNAs are generated by PCR, primer pairs suitable to amplify DNA templates from genomic DNA or cDNA are calculated. In addition, E-RNAi allows the access of pre-designed dsRNAs from published experiments.
- GeneScript siRNA Calculator v1.0 beta (http://proteas.uio.no/siRNA beta.html) scans an input cDNA sequence derived from a human gene and applies some sequence pattern rules. No specificity check is performed.
- GeneScript siRNA Construct Builder (https://www.genscript.com/ssl-bin/app/rnai?op=known) (Wang and Mu, 2004) builds shRNAs from siRNAs. They also have a tool to build scrambled siRNA controls.

- GPboost (Saetrom, 2004) is a genetic programming-based prediction system using weighted sum of sequence motifs/patterns. The prediction system uses proprietary hardware and is available for both commercial and strategic academic collaborations. Their siRNA database is available upon request.
- OligoEngine (http://www.oligoengine.com/index.html) (Brummelkamp et al., 2002) can create shRNA and siRNA, predicts secondary structures using Mfold and BLAST against UniGene. A standalone version is available. Both the web and standalone version require registration.
- OligoFactory (http://ueg.ulb.ac.be/oligofaktory/) (Schretter and Milinkovitch, 2006) uses the rules developed by Reynolds et al. (2004).
- RNAi Central (http://katahdin.cshl.org:9331/portal/scripts/main2.pl) can predict siRNAs, shRNAs, 29mer overhangs among others and is based on the rules used by Paddison et al. (2002) and Silva et al. (2005). The specificity is checked for human, mouse and others.
- RNAi Explorer (http://www.genelink.com/sirna/shRNAi.asp) searches the open reading frame (ORF) and uses GC content and basic sequence patterns to select candidate shRNAs that can be BLASTed using the NCBI web services.
- RNA Oligo Retriever (http://katahdin.cshl.org:9331/RNAi/html/rnai.html) (Paddison et al., 2002) designs siRNAs for different expression systems and applies position-based rules. The sequence of the passenger strand can be modified by the program to achieve better specificity and knockdown effects.
- SciTools RNAi Design (http://www.idtdna.com/Scitools/Applications/RNAi/RNAi.aspx) (Kim et al., 2005) can generate 21mer and 27mer siRNAs using the rules from Elbashir et al. (2001a, 2001b) or their specific unified rule set. Individual BLAST searches are available.
- SDS (siRNA Design Software) (http://i.cs.hku.hk/~sirna/software/sirna.php) (Yiu et al., 2005) makes use of existing design tools to output a set of candidates. The candidates are then filtered based on their secondary structure prediction.
- SiDE (http://side.bioinfo.cipf.es/) (Santoyo et al., 2005) implements filters based on sequence pattern, SNPs, exon boundaries to identify siRNA sequences, and performs BLAST searches. The sequences cannot be entered directly but only by using identifiers of common sequence databases.
- siDESIGN Center (http://design.dharmacon.com/) builds on early guidelines by Elbashir et al. (2001a, 2001b) and adds eight additional criteria described by Reynolds and colleagues (2004). This program offers the flexibility of defining specific target regions, adjusting certain design criteria, and selecting BLAST. Ranked lists of candidate siRNA sequences are provided along with siRNA sequences for all designs performed.
- Sirna (http://sfold.wadsworth.org/sirna.pl) is a specialized tool for target accessibility prediction and RNA duplex thermodynamics for the rational siRNA design. It is based on an advanced thermodynamics algorithm (Ding and Lawrence, 2003). The program does not perform any specificity check.

- SiRnaDesigner (http://www1.qiagen.com/Products/GeneSilencing/Cus
  tomSiRna/SiRnaDesigner.aspx) (Huesken *et al.*, 2005) uses rules by
  Elbashir *et al.* (2001a, 2001b) with additional parameters for (i) differ-
  ential melting temperature of the 5′- and 3′-ends; (ii) overall GC
  content; (iii) base preferences at specific sites; and (iv) avoidance of
  stretches of Gs or Cs. Sequences can be individually BLASTed.
- siRNA scales (http://gesteland.genetics.utah.edu/siRNA_scales/) calculates two parameters: (i) stability between 5′ antisense and 5′ sense
  duplex ends as an asymmetry parameter using dinucleotide energies;
  and (ii) the total GC content.
- SiRNA selection Demo (Interagon: https://demo1.interagon.com/
  demo/) (Snove *et al.*, 2004) requires registration. It offers some of the
  published algorithms as separate filters, which include Saetrom (2004),
  Amarzguioui and Prydz (2004), Chalk *et al.* (2004) and others. The
  demo version uses mouse cDNA from Ensembl's m34 release of
  December 2005 for specificity checks.
- siRNA selection Program at the Whitehead Institute (http://
  jura.wi.mit.edu/siRNAext) (Yuan *et al.*, 2004) applies pattern matching
  rules (Elbashir *et al.*, 2001b) and calculates some binding energy. The
  specificity is checked against human, mouse, and rat.
- siRNA selector (http://hydra1.wistar.upenn.edu/Projects/siRNA/siRNA
  index.htm) (Levenkova *et al.*, 2004). A set of rules is used for evaluating
  siRNA functionality based on thermodynamics parameters (Khvorova *et
  al.*, 2003; Schwarz *et al.*, 2003) and sequence-related determinants
  developed by Dharmacon (Reynolds *et al.*, 2004). The specificity is
  determined using BLAST against UniGene databases.
- siRNA Target Finder (https://www.genscript.com/ssl-bin/app/rnai)
  (Wang and Mu, 2004) identifies unique siRNA target sequences based
  on thermodynamic properties, RNA secondary structure predictions of
  the siRNA, immune response predictions (Hornung *et al.*, 2005; Judge *et
  al.*, 2005), and other parameters.
- siSearch (http://sonnhammer.cgb.ki.se/siSearch/siSearch_1.7.html)
  (Chalk *et al.*, 2004) is designed to select siRNAs with rules derived from
  various sources, including those from Amarzguioui and Prydz (2004),
  Chalk *et al.* (2004), Jagla *et al.* (2005), Reynolds *et al.* (2004), and Ui-Tei
  et al. (2004), calculates energy conditions and checks for specificity. For
  this purpose, Unigene and Refseq can be used as a reference database.
- SVM RNAi (http://www.changbioscience.com/stat/sirna.html). Version
  2.0 is available online. Version 3.6 is available as a trial and can be
  purchased. It uses a trained support vector machine classifier to predict
  siRNAs.
- TROD (http://www.unige.ch/sciences/biologie/bicel/websoft/RNAi
  .html) (Dudek and Picard, 2004) scans for appropriate target sequences
  based on the constraints of the T7 RNA pol method and published
  criteria for RNA interference with siRNAs.

2.4.1 Other related tools

- miRU (http://bioinfo3.noble.org/miRNA/miRU.htm) (Zhang, 2005)
  predicts plant miRNA target genes. It reports all potential sequences

complementary to the query with mismatches no more than specified for each mismatch type. Mismatch types distinguish between G:U Wobble pairs, indels and other mismatches. In addition, each mismatch is penalized according to the mismatch type and position in the miRNA.

2.5 Databases

Databases that provide sequences for already tested and verified siRNA sequences can be a great resource and increase the success rate drastically. However, choosing an siRNA sequence from these databases needs caution because these sequences have been tested in a specific organism or cell type under specific conditions. Modifying those conditions can potentially change the cell response by, for example, increasing the chance of off-target effect. Thus, it will still be necessary to test more than one sequence. Other types of databases provide access to cloned shRNAs, or predicted siRNAs, which are not necessarily experimentally tested. The following links refer to existing siRNA and miRNA databases available online as of May 2006.

2.5.1 siRNA databases

- RNAi Database (http://rnai.org or http://nematoda.bio.nyu.edu/cgi-bin/rnai/index_col2.cgi). This database contains worm RNAi data, including assay, resulting phenotypes, dsRNA target location.
- RNAi Phenotypes (http://www.wormbase.org/db/searches/rnai_search) is a collection of RNAi experimental data from *C. elegans*.
- siRNA Database (http://www.proteinlounge.com/sirna_home.asp) – (for fee access). This database provides pre-generated, not experimentally tested, siRNA sequences against all known genes for a variety of organisms.
- siRNA Resource at CGB, KI (http://gemini.cgb.ki.se:8080/sirnadb/index.jsp). There are currently 1276 experimentally validated siRNAs targeting 635 genes in the database. It also contains 64 166 predicted siRNAs. The current release is 1.0, and the last update is 20 September 2005.
- The RNAi Consortium shRNA Library (The RNAi Consortium; TRC) (http://www.broad.mit.edu/genome_bio/trc/rnai.html) contains shRNA clones produced by the TRC, as well as protocols for handling and conducting screens with shRNA molecules. The RNAi consortium shRNA library is distributed as bacterial glycerol stocks, plasmid DNA or lentiviral particles by Sigma-Aldrich and as bacterial glycerol stocks by Open Biosystems.
- siRecords (http://siRecords.umn.edu/siRecords/) (Ren *et al.*, 2006) is a database of known functional siRNAs and includes literature information, cell type, sequence information, efficacy classification among other useful information.
- Ambion's siRNA database (http://www.ambion.com/catalog/sirna_search.php) contains a list of verified siRNAs.
- The DSTHO database (http://203.199.182.73/gnsmmg/databases/sirna/dstho.html) (Dash *et al.*, 2006) lists siRNA for Oncogenes.

- HuSiDa (http://itb.biologie.hu-berlin.de/~nebulus/sirna/v2/) (Truss *et al.*, 2005) is a public database that serves as a depository for both sequences of published functional siRNA molecules targeting human genes and important technical details of the corresponding gene silencing experiments.
- RNA interference (http://rnainterference.org/Sequences.html) provides lists of used siRNAs with Pubmed references for human, mouse, rat, and other organisms.

2.5.2 miRNA databases

- http://www.microrna.org at Memorial Sloan Kettering Cancer Center lists identified miRNA in human, drosophila and zebra fish and includes a prediction tool for miRNA (miRanda).
- RNAi Codex (http://codex.cshl.edu/scripts/newmain.pl). Presently, the Codex database describes clones from the Hannon–Elledge shRNA libraries (mouse and human) that are available through Open Biosystems.

2.6 Target RNA secondary structure predictions

The local mRNA target structure should be accessible to RISC for efficient targeting. The local structure can, to some extent, be predicted by computer programs, and it is therefore advisable to check the target structure with these programs. Target accessibility has long been established as an important factor for the potency of antisense oligonucleotides and *trans*-cleaving ribozymes. Recently, the importance of target structure and accessibility in determining the potency of siRNAs has been demonstrated, using a number of experimental approaches that include oligo library (Lee *et al.*, 2002), oligo array (Bohula *et al.*, 2003), antisense evaluation of accessibility (Kretschmer-Kazemi Far and Sczakiel, 2003), and by targeting the same sequence in both structured and unstructured sites (Vickers *et al.*, 2003).

Using mRNAs with engineered secondary structures, Schubert *et al.* (2005) showed that structural features of the target mRNA have a significant effect on siRNA activity. They showed that the activity of a highly active siRNA can be drastically diminished when target nucleotides are incorporated into various hairpin structures. There is also a linear correlation between the siRNA efficacy and the local free energy, predicted using Mfold (Mathews *et al.*, 1999b; Zuker, 1989). In addition, Luo and Chang found that the number of hydrogen bonds that are formed between the target region and the rest of the mRNA is a useful parameter, correlating negatively to silencing efficiency (Luo and Chang, 2004). Patzel and coworkers (2005) showed that siRNAs with more terminal free nucleotides, especially at the 3'-end, are more active than others. Using secondary structure prediction of the guide strand, a correlation between its predicted secondary structure and activity has also been identified (Mathews *et al.*, 1999b; Zuker, 1989).

Unfortunately, it is generally very difficult to correctly model the complex secondary structure of mRNAs (Lima *et al.*, 1992; Overhoff *et al.*, 2005; Sohail *et al.*, 1999; Stein, 2001). Free energy minimization prediction of mRNA structures only predicts about 70% of the known canonical base pairs (Mathews *et al.*, 1999b). Local secondary structure predictions are complicated because there might be many local structures that have similar energies and are regarded equal when calculating global structures. Mfold, for example, can give many different structures all with similar energy values. Using this ensemble of structures or picking the right one is, therefore, a tremendous task.

Current approaches to optimize the prediction accuracy for local mRNA structures try to address this problem (Heale *et al.*, 2005; Ogurtsov *et al.*, 2006), others include using microarray experiments (Kierzek *et al.*, 2006) or optimizing established algorithms for the task of precise local structure prediction (Ding *et al.*, 2004; Muckstein *et al.*, 2006b). Heale *et al.* (2005) used Mfold and analyzed folding results with respect to local folding features. They could improve the percentage of correctly predicted siRNAs by about 10%.

The following is a list of web resources on mRNA folding:

- RNALOSS (Clote, 2005) (http://clavius.bc.edu/~clotelab/RNALOSS) estimate if an RNA of up to 100 nucleotides is highly structured according to Boltzmann probability.
- RNAstructure 4.3 (Mathews *et al.*, 1999a) (http://rna.urmc.rochester. edu/rnastructure.html) is an MS Windows program for the prediction and analysis of RNA secondary structures. It includes OligoWalk, which predicts the equilibrium affinity of complementary DNA or RNA oligonucleotides to an RNA target.
- Mfold (Mathews *et al.*, 1999b; Zuker, 1989, 2003) (http://www. bioinfo.rpi.edu/applications/mfold/) is one of the first and most widely used program to predict secondary structures of RNA.
- Vienna RNA Package (Hofacker, 2003) (http://rna.tbi.univie.ac.at/). The standalone version of this program predicts standard secondary structure.
- Sfold web server (http://www.bioinfo.rpi.edu/applications/sfold/ sirna.pl) (Ding *et al.*, 2004). This algorithm generates a statistical sample of RNA secondary structures from the Boltzmann ensemble of RNA secondary structures.
- RNAup (Vienna package) (Muckstein *et al.*, 2006a, 2006b; Schuster *et al.*, 1994) (http://www.tbi.univie.ac.at/~ulim/RNAup/). RNAup is a standalone program that calculates the thermodynamics of RNA–RNA interactions. RNA–RNA binding is decomposed into two stages: (i) first the partition function for secondary structures of the target RNA is computed, which is subject to the constraint that a certain sequence interval (the binding site) remains unpaired; (2) then, the binding energy in the target is calculated as the optimum over all possible types of bindings.
- Afold (Ogurtsov *et al.*, 2006) (ftp://ftp.ncbi.nlm.nih.gov/pub/ogurtsov/ Afold) predicts the optimal secondary structure of RNA molecules of up to 28 000 nucleotides.
- MSARI (http://theory.csail.mit.edu/MSARi) (Coventry *et al.*, 2004) is a program for detecting conservation of RNA secondary structure. It searches orthologous nucleotide sequences for statistically significant variations conserving a candidate secondary structure.

- Kinefold (http://kinefold.curie.fr/) (Xayaphoummine *et al.*, 2005) is a program for stochastic folding simulations of nucleic acids on second to minute molecular time scales.

2.7 Other resources on the web

Here are additional useful resources on the World Wide Web that present some interest to the researcher who wants to use the RNAi technology:

- Google's web page on RNAi http://www.google.com/Top/Science/Biology/Biochemistry_and_Molecular_Biology/Gene_Expression/RNA_Interference/
- http://rnainterference.org/ is a database that contains many resources on RNAi studies.
- http://www.ambion.com/techlib/tb/tb_506.html describes some technical aspects of RNAi.
- http://www.rnaiweb.com/ is a resource center for RNAi technology.
- http://www.stz-nad.com/main.html (Steinbeis Transfer Center for Nucleic Acids Design). This web page was still under construction during preparation of this manuscript, but promised to provide relevant tools for the design of siRNAs among others.
- http://www.rockefeller.edu/labheads/tuschl/sirna.html, Tuschl laboratory web site.
- http://bioinformatics.ubc.ca/resources/links_directory (Fox *et al.*, 2005). The Bioinformatics Links Directory features curated links to molecular resources, tools, and databases.
- http://www.bioinformatics.ubc.ca/resources/links_directory/index.php?search=siRNA provides a directory of bioinformatics link.
- http://www.openbiosystems.com/rnai/ gives information about mammalian, non-mammalian libraries, cloning vectors, supporting products like controls, transfection reagents, etc.
- http://gesteland.genetics.utah.edu/members/olgaM/otherSW.html provides some collections of datasets and tools including siRNA scales.

2.8 Concluding remarks

RNA interference has become the leading methodology for gene knock-down experiments. To successfully use RNAi, several factors have to be taken into consideration. When designing RNAi experiments we advise the following tests be performed:

(i)   Check if there are already experimentally tested siRNAs for the gene of interest.
(ii) If no siRNA for the given purpose is available, use one of the advanced publicly available design tools like BIOPREDsi, OligoEngine, or RNAi Central to predict siRNAs.
(iii) Check that the target region is not likely to be in a secondary structure.
(iv) Make sure that the siRNA is highly asymmetric and is not likely to form secondary structures.
(v)   Compare the target sequence and the siRNAs with the genome of interest.
(vi) Use more than two siRNAs to verify your results.

Following these guidelines will provide you with siRNAs with the highest probability of success based on our current knowledge.

Acknowledgments

We apologize to the many researchers in the RNA interference field whose work was not mentioned in this manuscript. We thank Peng Jiang from Southeast University, China, Xiaowei Wang (Ambion Inc.), and Olga Matveeva (University of Utah) for helpful discussions. We are grateful to Drs Matthew Beard, Allen Volchuk, and Ai Yamamoto for helpful discussions and reading the manuscript.

References

Akeju, O., Peng, T. and Park, J.K. (2006) Short hairpin RNA loop design for the facilitation of sequence verification. *Biotechniques* **40:** 154, 156, 158.

Altschul, S.F., Gish, W., Miller, W., Myers, E.W. and Lipman, D.J. (1990) Basic local alignment search tool. *J Mol Biol* **215:** 403–410.

Amarzguioui, M. and Prydz, H. (2004) An algorithm for selection of functional siRNA sequences. *Biochem Biophys Res Commun* **316:** 1050–1058.

Amarzguioui, M., Rossi, J.J. and Kim, D. (2005) Approaches for chemically synthesized siRNA and vector-mediated RNAi. *FEBS Letters* **579:** 5974–5981.

Ambros, V. (2001) microRNAs: tiny regulators with great potential. *Cell* **107:** 823–826.

Aronin, N. (2006) Target selectivity in mRNA silencing. *Gene Therapy* **13:** 509–516.

Arziman, Z., Horn, T. and Boutros, M. (2005) E-RNAi: a web application to design optimized RNAi constructs. *Nucleic Acids Res* **33:** W582–W588.

Berns, K., Hijmans, E.M., Mullenders, J., *et al.* (2004) A large-scale RNAi screen in human cells identifies new components of the p53 pathway. *Nature* **428:** 431–437.

Birmingham, A., Anderson, E.M., Reynolds, A., *et al.* (2006) 3′ UTR seed matches, but not overall identity, are associated with RNAi off-targets. *Nat Methods* **3:** 199–204.

Bohula, E.A., Slaisbury, A.J., Sohail, M., Playford, M.P., Riedemann, J., Southern, E.M. and Macaulay, V.M. (2003) The efficacy of small interfering RNAs targeted to the type 1 insulin-like growth factor receptor (IGF1R) is influenced by secondary structure in the IGF1R transcript. *J Biol Chem* **278:** 15991–15997.

Brummelkamp, T.R., Bernards, R. and Agami, R. (2002) A system for stable expression of short interfering RNAs in mammalian cells. *Science* **296:** 550–553.

Chalk, A.M., Wahlestedt, C. and Sonnhammer, E.L.L. (2004) Improved and automated prediction of effective siRNA. *Biochem Biophys Res Commun* **319:** 264–274.

Clote, P. (2005) RNALOSS: a web server for RNA locally optimal secondary structures. *Nucleic Acids Res* **33:** W600–W604.

Coventry, A., Kleitman, D.J. and Berger, B. (2004) MSARI: multiple sequence alignments for statistical detection of RNA secondary structure. *Proc Natl Acad Sci USA* **101:** 12102–12107.

Dash, R., Moharana, S.S., Reddy, A.S., Sastry, G.M. and Sastry, G.N. (2006) DSTHO: database of siRNAs targeted at human oncogenes: a statistical analysis. *Int J Biol Macromolecules* **38:** 65–69.

Ding, Y., Chan, C.Y. and Lawrence, C.E. (2004) Sfold web server for statistical folding and rational design of nucleic acids. *Nucleic Acids Res* **32:** W135–W141.

Ding, Y. and Lawrence, C.E. (2003) A statistical sampling algorithm for RNA secondary structure prediction. *Nucleic Acids Res* **31:** 7280–7301.

Doench, J.G. and Sharp, P.A. (2004) Specificity of microRNA target selection in translational repression. *Genes Devel* **18**: 504–511.

Du, Q., Thonberg, H., Wang, J., Wahlestedt, C. and Liang, Z.C. (2005) A systematic analysis of the silencing effects of an active siRNA at all single-nucleotide mismatched target sites. *Nucleic Acids Res* **33**: 1671–1677.

Ducat, D.C., Herrera, F.J. and Triezenberg, S.J. (2003) Overcoming obstacles in DNA sequencing of expression plasmids for short interfering RNAs. *Biotechniques* **34**: 1140–1142, 1144.

Dudek, P. and Picard, D. (2004) TROD: T7 RNAi Oligo Designer. *Nucleic Acids Res* **32**: W121–W123.

Elbashir, S.M., Harborth, J., Lendeckel, W., Yalcin, A., Weber, K. and Tuschl, T. (2001a) Duplexes of 21-nucleotide RNAs mediate RNA interference in cultured mammalian cells. *Nature* **411**: 494–498.

Elbashir, S.M., Lendeckel, W. and Tuschl, T. (2001b) RNA interference is mediated by 21- and 22-nucleotide RNAs. *Genes Dev* **15**: 188–200.

Filipowicz, W. (2005) RNAi: the nuts and bolts of the RISC machine. *Cell* **122**: 17–20.

Fox, J.A., Butland, S.L., McMillan, S., Campbell, G. and Ouellette, B.F. (2005) The Bioinformatics Links Directory: a compilation of molecular biology web servers. *Nucleic Acids Res* **33**: W3–W24.

Haley, B. and Zamore, P.D. (2004) Kinetic analysis of the RNAi enzyme complex. *Nat Struct Mol Biol* **11**: 599–606.

Heale, B.S.E., Soifer, H.S., Bowers, C. and Rossi, J.J. (2005) siRNA target site secondary structure predictions using local stable substructures. *Nucleic Acids Res* **33**: e30.

Henschel, A., Buchholz, F. and Habermann, B. (2004) DEQOR: a web-based tool for the design and quality control of siRNAs. *Nucleic Acids Res* **32**: W113–W120.

Hofacker, I.L. (2003) Vienna RNA secondary structure server. *Nucleic Acids Res* **31**: 3429–3431.

Hornung, V., Guenthner-Biller, M., Bourquin, C., *et al.* (2005) Sequence-specific potent induction of IFN-alpha by short interfering RNA in plasmacytoid dendritic cells through TLR7. *Nat Med* **11**: 263–270.

Hsieh, A.C., Bo, R.H., Manola, J., Vazquez, F., Bare, O., Khvorova, A., Scaringe, S. and Sellers, W.R. (2004) A library of siRNA duplexes targeting the phosphoinositide 3-kinase pathway: determinants of gene silencing for use in cell-based screens. *Nucleic Acids Res* **32**: 893–901.

Huesken, D., Lange, J., Mickanin, C., *et al.* (2005) Design of a genome-wide siRNA library using an artificial neural network. *Nat Biotechnol* **23**: 995–1001.

Huppi, K., Martin, S.E. and Caplen, N.J. (2005) Defining and assaying RNAi in mammalian cells. *Mol Cell* **17**: 1–10.

Hutvagner, G. (2005) Small RNA asymmetry in RNAi: function in RISC assembly and gene regulation. *FEBS Letters* **579**: 5850–5857.

Jagla, B., Aulner, N., Kelly, P.D., *et al.* (2005) Sequence characteristics of functional siRNAs. *RNA* **11**: 864–872.

Judge, A.D., Sood, V., Shaw, J.R., Fang, D., McClintock, K. and MacLachlan, I. (2005) Sequence-dependent stimulation of the mammalian innate immune response by synthetic siRNA. *Nat Biotechnol* **23**: 457–462.

Ketting, R.F., Fischer, S.E.J., Bernstein, E., Sijen, T., Hannon, G.J., Plasterk, R.H.A. (2001) Dicer functions in RNA interference and in synthesis of small RNA involved in developmental timing in *C. elegans*. *Genes Devel* **15**: 2654–2659.

Khvorova, A., Reynolds, A. and Jayasena, S.D. (2003) Functional siRNAs and rniRNAs exhibit strand bias. *Cell* **115**: 209–216.

Kierzek, E., Kierzek, R., Turner, D.H. and Catrina, I.E. (2006) Facilitating RNA structure prediction with microarrays. *Biochemistry* **45**: 581–593.

Kim, D.H., Behlke, M.A., Rose, S.D., Chang, M.S., Choi, S. and Rossi, J.J. (2005) Synthetic dsRNA Dicer substrates enhance RNAi potency and efficacy. *Nat Biotechnol* **23**: 222–226.

Kretschmer-Kazemi Far, R. and Sczakiel, G. (2003) The activity of siRNA in mammalian cells is related to structural target accessibility: a comparison with antisense oligonucleotides. *Nucleic Acids Res* **31**: 4417–4424.

Lee, N.S., Dohjima, T., Bauer, G., Li, H., Li, M.J., Ehsani, A., Salvaterra, P. and Rossi, J. (2002) Expression of small interfering RNAs targeted against HIV-1 rev transcripts in human cells. *Nat Biotechnol* **20**: 500–505.

Levenkova, N., Gu, Q.J. and Rux, J.J. (2004) Gene specific siRNA selector. *Bioinformatics* **20**: 430–432.

Lima, W.F., Monia, B.P., Ecker, D.J. and Freier, S.M. (1992) Implication of RNA structure on antisense oligonucleotide hybridization kinetics. *Biochemistry* **31**: 12055–12061.

Luo, K.Q. and Chang, D.C. (2004) The gene-silencing efficiency of siRNA is strongly dependent on the local structure of mRNA at the targeted region. *Biochem Biophys Res Commun* **318**: 303–310.

Ma, J.B., Yuan, Y.R., Meister, G., Pei, Y., Tuschl, T. and Patel, D.J. (2005) Structural basis for 5′-end-specific recognition of guide RNA by the *A. fulgidus* Piwi protein. *Nature* **434**: 666–670.

Martinez, J., Patkaniowska, A., Urlaub, H., Luhrmann, R. and Tuschl, T. (2002) Single-stranded antisense siRNAs guide target RNA cleavage in RNAi. *Cell* **110**: 563–574.

Mathews, D.H., Burkard, M.E., Freier, S.M., Wyatt, J.R. and Turner, D.H. (1999a) Predicting oligonucleotide affinity to nucleic acid targets. *RNA* **5**: 1458–1469.

Mathews, D.H., Sabina, J., Zuker, M. and Turner, D.H. (1999b) Expanded sequence dependence of thermodynamic parameters improves prediction of RNA secondary structure. *J Mol Biol* **288**: 911–940.

Miyagishi, M., Matsumoto, S. and Taira, K. (2004) Generation of an shRNAi expression library against the whole human transcripts. *Virus Res* **102**: 117–124.

Muckstein, U., Tafer, H., Hackermuller, J., Bernhart, S.H., Stadler, P.F. and Hofacker, I.L. (2006a) Thermodynamics of RNA–RNA binding. *Bioinformatics* **22**: 1177–1182.

Muckstein, U., Tafer, H., Hackermuller, J., Bernhart, S.H., Stadler, P.F. and Hofacker, I.L. (2006b) Thermodynamics of RNA-RNA Binding. *Bioinformatics* btl024.

Ogurtsov, A.Y., Shabalina, S.A., Kondrashov, A.S. and Roytberg, M.A. (2006) Analysis of internal loops within the RNA secondary structure in almost quadratic time. *Bioinformatics* btl083.

Overhoff, M., Alken, M., Far, R.K.K., Lemaitre, M., Lebleu, B., Sczakiel, G. and Robbins, I. (2005) Local RNA target structure influences siRNA efficacy: a systematic global analysis. *J Mol Biol* **348**: 871–881.

Paddison, P.J., Caudy, A.A., Bernstein, E., Hannon, G.J. and Conklin, D.S. (2002) Short hairpin RNAs (shRNAs) induce sequence-specific silencing in mammalian cells. *Genes Devel* **16**: 948–958.

Paddison, P.J., Cleary, M., Silva, J.M., Chang, K., Sheth, N., Sachidanandam, R. and Hannon, G.J. (2004a) Cloning of short hairpin RNAs for gene knockdown in mammalian cells. *Nat Methods* **1**: 163–167.

Paddison, P.J., Silva, J.M., Conklin, D.S., *et al.* (2004b) A resource for large-scale RNA-interference-based screens in mammals. *Nature* **428**: 427–431.

Parker, J.S., Roe, S.M. and Barford, D. (2004) Crystal structure of a PIWI protein suggests mechanisms for siRNA recognition and slicer activity. *Embo J* **23**: 4727–4737.

Parker, J.S., Roe, S.M. and Barford, D. (2005) Structural insights into mRNA recognition from a PIWI domain-siRNA guide complex. *Nature* **434**: 663–666.

Patzel, V., Rutz, S., Dietrich, I., Koberle, C., Scheffold, A. and Kaufmann, S.H.E. (2005) Design of siRNAs producing unstructured guide-RNAs results in improved RNA interference efficiency. *Nat Biotechnol* **23**: 1440–1444.

Preall, J.B., He, Z.Y., Gorra, J.M. and Sontheimer, E.J. (2006) Short interfering RNA strand selection is independent of dsRNA processing polarity during RNAi in *Drosophila*. *Curr Biol* **16**: 530–535.

Ren, Y., Gong, W., Xu, Q., Zheng, X., Lin, D., Wang, Y. and Li, T. (2006) siRecords: an extensive database of mammalian siRNAs with efficacy ratings. *Bioinformatics* **22**: 1027–1028.

Reynolds, A., Leake, D., Boese, Q., Scaringe, S., Marshall, W.S. and Khvorova, A. (2004) Rational siRNA design for RNA interference. *Nat Biotechnol* **22**: 326–330.

Saetrom, P. (2004) Predicting the efficacy of short oligonucleotides in antisense and RNAi experiments with boosted genetic programming. *Bioinformatics* **20**: 3055–3063.

Santoyo, J., Vaquerizas, J.M. and Dopazo, J. (2005) Highly specific and accurate selection of siRNAs for high-throughput functional assays. *Bioinformatics* **21**: 1376–1382.

Schretter, C. and Milinkovitch, M.C. (2006) OLIGOFAKTORY: a visual tool for interactive oligonucleotide design. *Bioinformatics* **22**: 115–116.

Schubert, S., Grunweller, A., Erdmann, V.A. and Kurreck, J. (2005) Local RNA target structure influences siRNA efficacy: systematic analysis of intentionally designed binding regions. *J Mol Biol* **348**: 883–893.

Schuster, P., Fontana, W., Stadler, P.F. and Hofacker, I.L. (1994) From sequences to shapes and back: a case study in RNA secondary structures. *Proc Biol Sci* **255**: 279–284.

Schwarz, D.S., Hutvagner, G., Du, T., Xu, Z.S., Aronin, N. and Zamore, P.D. (2003) Asymmetry in the assembly of the RNAi enzyme complex. *Cell* **115**: 199–208.

Silva, J.M., Li, M.Z., Chang, K., *et al.* (2005) Second-generation shRNA libraries covering the mouse and human genomes. *Nat Genet* **37**: 1281–1288.

Siolas, D., Lerner, C., Burchard, J., Ge, W., Linsley, P.S., Paddison, P.J., Hannon, G.J. and Cleary, M.A. (2005) Synthetic shRNAs as potent RNAi triggers. *Nat Biotechnol* **23**: 227–231.

Snove, O., Nedland, M., Fjeldstad, S.H., Humberset, H., Birkeland, O.R., Grunfeld, T. and Saetrom, P. (2004) Designing effective siRNAs with off target control. *Biochem Biophys Res Commun* **325**: 769–773.

Sohail, M., Akhtar, S. and Southern, E.M. (1999) The folding of large RNAs studied by hybridization to arrays of complementary oligonucleotides. *RNA* **5**: 646–655.

Stein, C.A. (2001) The experimental use of antisense oligonucleotides: a guide for the perplexed. *J Clin Invest* **108**: 641–644.

Taxman, D.J., Livingstone, L.R., Zhang, J.H., Conti, B.J., Iocca, H.A., Williams, K.L., Lich, J.D., Ting, J.P.Y. and Reed, W. (2006) Criteria for effective design, construction, and gene knockdown by shRNA vectors. *BMC Biotechnol* **6**: 7.

Truss, M., Swat, M., Kielbasa, S.M., Schafer, R., Herzel, H. and Hagemeier, C. (2005) HuSiDa – the human siRNA database: an open-access database for published functional siRNA sequences and technical details of efficient transfer into recipient cells. *Nucleic Acids Res* **33**: D108–D111.

Ui-Tei, K., Naito, Y., Takahishi, F., Haraguchi, T., Ohki-Hamazaki, H., Juni, A., Ueda, R. and Saigo, K. (2004) Guidelines for the selection of highly effective siRNA sequences for mammalian and chick RNA interference. *Nucleic Acids Research* **32**: 936–948.

Vickers, T.A., Koo, S., Bennett, C.F., Crooke, S.T., Dean, N.M. and Baker, B.F. (2003) Efficient reduction of target RNAs by small interfering RNA and RNase H-dependent antisense agents. A comparative analysis. *J Biol Chem* **278**: 7108–7118.

Wang, L. and Mu, F.Y. (2004) A Web-based design center for vector-based siRNA and siRNA cassette. *Bioinformatics* **20**: 1818–1820.

Xayaphoummine, A., Bucher, T. and Isambert, H. (2005) Kinefold web server for RNA/DNA folding path and structure prediction including pseudoknots and knots. *Nucleic Acids Res* **33**: W605–W610.

Yiu, S.M., Wong, P.W.H., Lam, T.W., Mui, Y.C., Kung, H.F., Lin, M. and Cheung, Y.T. (2005) Filtering of ineffective siRNAs and improved siRNA design tool. *Bioinformatics* **21**: 144–151.

Yoshinari, K., Miyagishi, M. and Taira, K. (2004) Effects on RNAi of the tight structure, sequence and position of the targeted region. *Nucleic Acids Res* **32**: 691–699.

Yuan, B., Latek, R., Hossbach, M., Tuschl, T. and Lewitter, F. (2004) siRNA Selection Server: an automated siRNA oligonucleotide prediction server. *Nucleic Acids Res* **32**: W130–W134.

Yuan, Y.R., Pei, Y., Ma, J.B., *et al.* (2005) Crystal structure of *A. aeolicus* Argonaute, a site-specific DNA-guided endoribonuclease, provides insights into RISC-mediated mRNA cleavage. *Mol Cell* **19**: 405–419.

Zeng, Y., Wagner, E.J. and Cullen, B.R. (2002) Both natural and designed micro RNAs technique can inhibit the expression of cognate mRNAs when expressed in human cells. *Mol Cell* **9**: 1327–1333.

Zhang, Y. (2005) miRU: an automated plant miRNA target prediction server. *Nucleic Acids Res* **33**: W701–W704.

Zuker, M. (1989) On finding all suboptimal foldings of an RNA molecule. *Science* **244**: 48–52.

Zuker, M. (2003) Mfold web server for nucleic acid folding and hybridization prediction. *Nucleic Acids Res* **31**: 3406–3415.

RNAi – a chemical perspective

3

Ouathek Ouerfelli

3.1 Background

The discovery of RNAi (Fire *et al.*, 1998) over the past few years has taken the biological and medical sciences by surprise (for reviews and guidance, see Couzin, 2002; Hammond *et al.*, 2000; Hannon, 2002; Nykaneri *et al.*, 2001; Sharp, 2001; Tuschl, 2002; Zamore, 2001; Zamore *et al.*, 2000). The recent awarding of the 2006 Nobel Prize to the RNAi field is an early indication of its promise. After a short infancy as a post-transcriptional modification in *C. elegans* (Fire *et al.*, 1998), as well as a quelling technique in plants (Jorgensen, 1990; Romano and Macino, 1992), it emerged as a powerful ubiquitous knockdown method that has quickly moved to vertebrates including human studies (Cullen, 2006). Ever since Tuschl and colleagues, in their milestone discovery in 2001 using chemically synthesized RNA duplexes (Caplen *et al.*, 2001; Elbashir *et al.*, 2001a, 2001b), uncovered the size exclusion secret that has precluded RNAi use in humans, the technique has shown a great deal of promise in gene validation, as a therapeutic (Behlke, 2006; Sachse and Echeverri, 2004; Sandy *et al.*, 2005; Smith, 2006) and especially as an anti-infective agent (Novina *et al.*, 2002; Qin *et al.*, 2003; Wilson *et al.*, 2003; Zamore and Aronin, 2003). Indeed, 21–27-long RNA duplexes with two-nucleotide overhang (Elbashir *et al.*, 2001c) could bypass the immunological guards at more than 30 nucleotides to undergo RISC processing and antisense-strand-guided damage of the respective complementary mRNA (Martinez *et al.*, 2002; Schwarz *et al.*, 2002). Furthermore, the combination of RNAi discovery with the recent sequencing of several genomes, including human, has made the global study of gene function within a given genome in a living cell suddenly an attainable goal.

As in any newly discovered discipline, challenges abound. At the sequence choice level, not all siRNAs are equally active. At the transfection stage, besides the intrinsic toxicity of each transfection reagent, every cell line or kind needs its own transfection protocol. Finally, providing needed siRNAs in genome-scale numbers with several factors such as annotation, and siRNA design algorithms, in continuous improvement has added complexity to the task.

In this chapter, an overview of the chemical contribution to RNAi will be presented. Special focus will be given to chemical strategies that provided high-throughput synthesis of modified and unmodified siRNAs.

Another area where chemistry is also playing an increasing role is chemical transfection of siRNAs into cells (Hogrefe *et al.*, 2006; Spagnou *et al.*, 2004; Utku *et al.*, 2006). This is another area that needs special attention. Indeed, some of the toxicities and off-target effects have been attributed to the transfection reagent itself (Scherer and Rossi, 2004). The fact that there is no single transfection reagent for all cells is indicative of the challenges ahead.

3.2 Introduction

One of the greatest advantages that RNAi has brought is the ability to screen living cells and harness phenotypic as well as biochemical changes consequent to loss of function. To be able to mine whole genomes, many technologies have to be developed and adapted to high-content, high-throughput screening of living cells. Assay development, readouts, miniaturization, bioinformatics, and chemical support had to respond to the exponential demand. While most biological sciences could quickly adjust to the neo-situation by redirecting existing technologies, some other sciences had yet to answer to the new challenge. Morphological and phenotypic readouts are being sought to address genome-wide screens that ideally require advanced automated microscopy, pattern recognition, a great deal of automation, and reliable integrated technologies that minimize human intervention while maximizing readout accuracy and decision making.

Over the past year, there have been major breakthroughs in siRNA design that maximized siRNA potency (Birmingham *et al.*, 2006; Jagla *et al.*, 2005; Scherer and Rossi, 2004) while alleviating if not eliminating off-target effects (Chi *et al.*, 2003; Jakson *et al.*, 2003; Scherer and Rossi, 2004; Semizarov *et al.*, 2003). Several genome-wide siRNA screens were published that pointed out additional challenges as well as validation of the field (Czauderna *et al.*, 2003; Moffat and Sabatini, 2006; Scherer and Rossi, 2006). On the chemical side, there were many contributions by very few laboratories mainly in industry. This chapter will give a quick overview of the chemical response to the challenge and will also point out some of the opportunities to address.

The chemical intervention is sought to provide a quick response and flexibility to the moving target that is the design algorithm. Chemistry could address the new promise of siRNAs as therapeutic agents. This has brought into play the problem of stabilization of the duplexes as well as the antisense strand itself towards nucleases (Czauderna *et al.*, 2003; Moffat and Sabatini, 2006; Scherer and Rossi, 2006), and finding delivery vehicles (Hogrefe *et al.*, 2006; Spagnou *et al.*, 2004; Utku *et al.*, 2006).

3.3 siRNAs versus shRNAs

siRNAs can be chemically synthesized or expressed. The latter are commonly referred to as shRNAs (Brummelkamp *et al.*, 2002; Lee *et al.*, 2002; McManus and Sharp, 2002; McManus *et al.*, 2002; Paddison *et al.*, 2002). Both reagents are nowadays available from many sources, even plated for high-throughput screening. However, each strategy has its advantages and disadvantages.

3.3.1 siRNAs

Chemically synthesized siRNAs could be synthesized in large amounts and numbers (see below), and are amenable to chemical modifications for specific needs. Modifications could be introduced for increased stability (Amarzguioui *et al.*, 2003; Chiu and Rana, 2002; Holen *et al.*, 2003; Scherer and Rossi, 2004), monitoring, membrane permeability, and in response to specific designs. This is especially important at this time, when transfecting siRNAs into cells is still a limiting factor. Notable additional advantages are the flexibility to screen them the same way that small molecule libraries are screened. They also offer the possibility to titrate a given effect to further validate a target.

3.3.2 shRNAs

shRNAs on the other hand, are inherently labor intensive, but for a given mRNA, should the respective protein be too stable for a knockdown to have any significance, or should the knockdown be intended for more than a few days, then conditional expression is the alternative.

From RNAi inception, it was clear that not every siRNA sequence is effective. This is mainly due to the difficulty in predicting RNA secondary structure and the difficulty in modeling it. As a consequence, scientists turned to pattern recognition as a means to compile factors that govern effectiveness based on sequences that worked and those that did not. Designing effective siRNAs is still a work in progress, and more and more clues are appearing regularly in the literature to enlighten this path (Amarzguioui *et al.*, 2003; Birmingham *et al.*, 2006; Chiu and Rana, 2002; Holen *et al.*, 2003; Jagla *et al.*, 2005; Scherer and Rossi, 2004). Another complication that is not clear to date is the fact that some sequences have the intrinsic ability to cause off-target effects, and some trigger interferon response. Although the latter could be alleviated by lowering siRNA concentration, the former is much more complicated. It could be that some sequences might target mRNA regions that are poorly defined. Ongoing studies will tell.

3.4 RNAi reagents

The best algorithms available recommend the use of three or more siRNAs per targeted gene. Given the 25 000–30 000 genes that compose the human genome, one needs in excess of a 100 000 siRNAs to be able to fully screen the human genome. Mouse and rat genomes will require bigger numbers. This has led some groups to target human and mouse genomes with one set of siRNAs. With the numbers being what they are, the next consideration is to choose the RNAi reagents: shRNAs or synthetic siRNAs, or a combination thereof. Using chemically synthesized siRNAs or expressed shRNAs is a very important point to consider, because of respective costs, and the advantages and disadvantages associated with each. These have been well reviewed recently (Behlke, 2006; Sachse and Echeverri, 2004; Sandy *et al.*, 2005; Smith, 2006). Suffice it to note that it is much easier to screen with siRNAs, but secondary screens with the aim of assessing most prolonged effects are better with shRNAs. While more expensive, siRNAs offer the possibility to titrate the effect of a knockdown, which is a very much sought-after advantage that is not possible with expressed shRNAs.

3.5 RNA chemistry

RNA chemistry has seen a great deal of advancement over the past decade. However, producing RNA duplexes in the hundreds of thousands, and at high speed will require much more accuracy, engineering, reliability, amenability to automation, and hopefully all of this is achieved while keeping cost as low as possible.

Automated solid phase RNA synthesis is conceptually identical to that of DNA synthesis. It involves a cycle of four steps: removal of the 5'-protecting group; adding the next phosphoramidite along with an activator; capping unreacted strands; and oxidation of the resulting phosphite triester to the corresponding phosphate (*Figure 3.1*). The difference between RNA and DNA synthesis is the presence of a protected 2'-hydroxyl group. This has a profound effect on stability, reactivity, as well as handing of the final sequences. In addition, RNA sequences are acid- and base-sensitive, and much more nuclease-sensitive than DNA, especially in the presence of bivalent ions like magnesium. The most common and most used protecting group is the tertiary-butyldimethylsilyl (TBDMS) group, which was introduced by Ogilvie *et al.* (1974). The TBDMS, as it served the good purpose of protecting the 2'OH and stabilized the sequence throughout the drastic synthesis steps, brought in steric hindrance that greatly influenced the reactivity of the phosphoramidite. TBDMS has also the propensity to shift to the 3'-position, which at times could complicate purification. Moreover, in DNA synthesis, coupling steps usually take place in a minute. However, in TBDMS-protected RNA, several repeat-coupling cycles (up to four or five) of 4 min are needed per nucleotide added. With RNA phosphoramidites being approximately 12 times more expensive than DNA ones, it was no surprise that scientists looked into improving the protecting group scheme to enhance yields and shorten coupling times, while saving valuable reagents.

3.6 RNA synthesis methods

Two recent major improvements have surfaced as the most reliable that were successfully applied in large-scale industrial settings. One was developed by the father of automated DNA synthesis – Marvin Caruthers and his group (Scaringe *et al.*, 1998). In their quest to find the mildest conditions in sequence purification post-synthesis and deblocking, they came up with a very original arrangement of protecting groups. They also found that a methyl phosphoramidite was more suitable than the cyanoethyl group for this particular strategy. This method is now commonly referred to as the 5'-O-SIL-2'-ACE method. To complete the picture, they modified existing automatic synthesizers to allow the use of their reagents. These are all available from Dharmacon (now part of Fisher Scientific).

The second major improvement was developed by Weiss and Pitsch (Pitsch *et al.*, 2001; Porcher and Pitsch, 2005), who came up with a less hindering substitute to the TBDMS – the triisopropylsilyoxymethyl (TOM) group. Besides its stability, its acetal nature has no propensity to shift to the

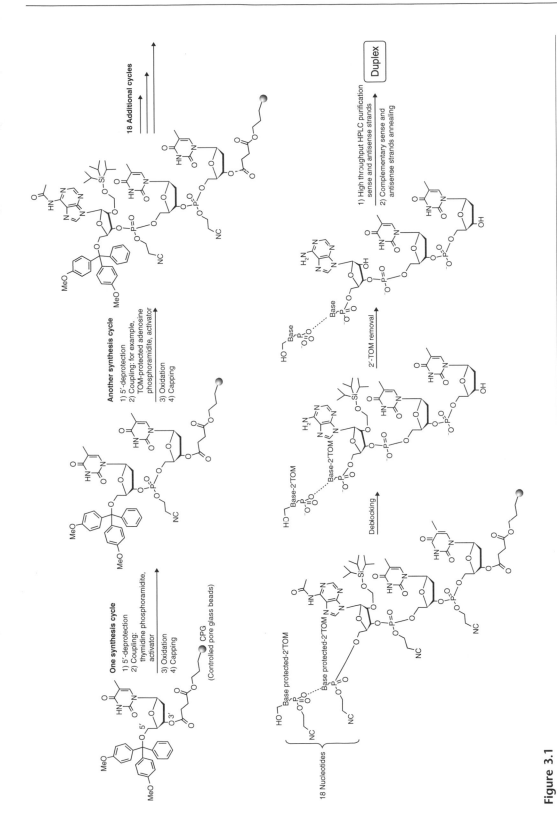

Figure 3.1

Automated synthesis of a single 21-nucleotide-long strand with two thymidines at the 3'-end, and additional processing that leads to duplex formation.

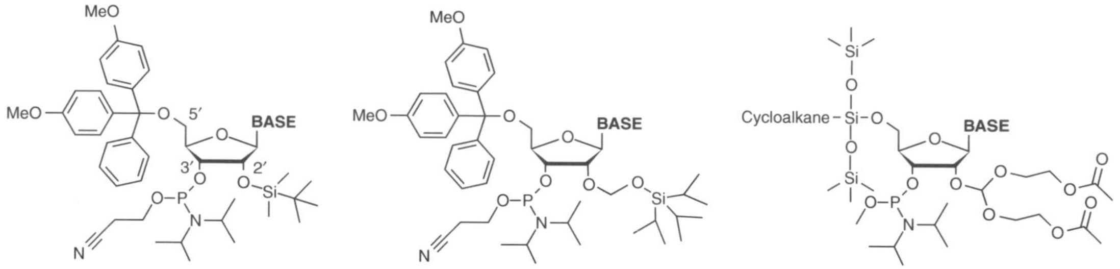

5'-DMT-2'-TBDMS ribonucleoside
phosphoramidites (Ogilvie)

5'-DMT-2'-TOM ribonucleoside
phosphoramidites (Weiss and Pitsch)

5'-O-SIL-2'-ACE ribonucleoside phosphoramidites
(Scaringe and Caruthers)

Figure 3.2

Widely used ribonucleoside phosphoramidites.

3'-position. 2'-TOM-protected phoshoramidites display reactivities that are closer to those of DNA phosphoramides than the TBDMS series. This methodology has also been widely used at an industrial scale and the TOM-phosphoramidites are now commercially available (*Figure 3.2*).

It goes without saying that although there are numerous new improvements that have appeared in the literature but have not reached wide applications, the more recent ones might (Efimov *et al.*, 2005; Muench and Pfleiderer, 2003; Ohgi *et al.*, 2005; Pon *et al.*, 2005; Reese, 2005).

Despite the above improvements in RNA synthesis, the issue of high-throughput strand purification, annealing, and quality control throughout the process remained to be addressed. Some siRNA providers prefer annealing and selling crude products. This might be harmless when the siRNA is highly active (more than 70% knockdown). In this case, whatever concentration gets in the cell will achieve the goal. However, at the genome-wide level, when all siRNAs cannot be validated, purification is the best approach.

3.7 High-throughput siRNA synthesis: the full process

In our laboratory (Ouerfelli *et al.*, manuscript in preparation), applying the TOM chemistry at high output sequences required access to large scale TOM-phosphoramidites. We have improved and scaled up TOM-phosphoramidite chemistry as a first step. In the second, we have tuned a DNA synthesizer to RNA synthesis. This was possible by working with manufacturers to provide us with special cartridges that allow slow gravitational passage of reagents through the solid support. By adjusting column packing, and packing consistency, to allow about 40 µl of reagent to flow through the synthesis column over a 4-min period, we could achieve high and reproducible yields.

Perhaps a highlight in our contribution was developing a better high-throughput purification means that is amenable to a production setting, especially with minimal additional steps. Known RNA purification methods relied on ion-exchange chromatography. Besides sequence dependence, elution of the final product involves high-salt solutions that have to be desalted in an additional step. We thought that it should be possible to achieve a short-cycle sequence-independent purification that allows the unattended purification through use of modern auto-inject, auto-collect HPLC. We have successfully applied and scaled up the quantitative purification of single strands through a combination of stationary support that is a mixed reverse phase-charge separation polymer, with a volatile buffer. This has permitted on average more than 95% purity of each single strand. In addition, we have made it possible to purify 192 strands per day.

In our efforts to produce high-purity duplexes, we have made it imperative to further check the quality of each single strand by applying high-throughput matrix-assisted laser desorption ionization–time of flight mass spectrometry (MALDI-TOF) against theoretical calculated values. Any sequences that failed were automatically resubmitted for synthesis.

The next additional hurdle to overcome was to devise an annealing process that permits a reproducible, human error-free concentration adjustment and annealing of complementary strands that is devoid of aggregated higher-order complexes. Single-strand synthesis yield is typically sequence-dependent. As a consequence, isolated amounts of sense and antisense sequences are different. Additionally, annealing requires the mixing of the equimolar amounts of the two complementary strands. Hence, following quantitative ultraviolet (UV) measurement of isolated amounts of each strand, the smaller of the two would be used in its entirety with equal stoichiometry of its complementary sequence. We have integrated the output of a 96-well UV spectrophotometer to a liquid handler to automatically adjust the concentrations of each strand up to a concentration that is double the final concentration of desired siRNA post-annealing. Furthermore, the liquid handler was programmed to mix equimolar amounts of the sense and antisense into respective pre-labeled vials on an annealing block.

Annealing is the process by which equimolar amounts of complementary sense and antisense strands are mixed and heated in annealing buffer up to an unstructured, disaggregated, and fully soluble state, then bringing down the temperature gradually to allow the two strands to anneal to each other without formation of unwanted side products. We have designed a 96-well aluminum block, and found optimum heating and cooling conditions that reproducibly yielded uniform duplexes as assessed by non-denaturing gel analyses. We have programmed the work-flow into a production. We have been able to provide 96 duplexes per day and a total of more than 22 000 siRNAs over a period of a year and half.

3.8 Summary and outlook

In conclusion, we have described most recent developments in high-throughput production of purified and annealed siRNAs. We have also taken the opportunity to give a bird's eye view of the steps required to chemically produce high-quality siRNAs in great numbers and innovative solutions that provided siRNAs in great numbers, and with high quality. This process could be easily multiplied to provide greater output. Although this was a 'first response' to a biological need, our method could be easily amenable to accommodate new designs and sequence requirements.

There is no doubt that chemistry will play an ever-increasing role in solving timely biological problems. As the RNAi field will ultimately gain more momentum as a therapy, chemistry should have solved all its *in vivo* stability and shelf life problems. Further chemical modifications of the ribose scaffold as well as the phosphate backbone could provide resistance to nucleases. Chemistry will also play a major role in keeping up with continuous siRNA design changes that improve specificity and reduce off-targeting. Finally, another area that is subject to intense chemical study at this time is siRNA transfection and delivery, and it is only a matter of time before it is unraveled.

Taken all together, the increasing collaboration between all biological sciences – medicine, physical sciences, and chemistry in particular – has begun to bear its fruits. The past 5 years have been very exciting for all involved in the field. Some of the challenges still persist, but the opportunities are also numerous. We have entered the era of personalized medicine and targeted treatments, and its tools are being prepared.

Acknowledgments

I am indebted to Drs James E. Rothman, Harold Varmus, Thomas Kelly, and Sir William M. Castell (then CEO of Amersham Biosciences) for advice and resources, Drs Dino De Angelis, Thomas Mayer, Bernd Jagla, Earl Kim, Da Song, Urs Rutishauser, Lee McDonald, Constantin Radu, Chunyan (Alice) Cao, Geralda Torchon, Andrzej Zatroski, Danuta Zatorska, Hui Fang, Minxue (Michelle) Liu, Nathalie Aulner, Allen Volchuk, Lars Branden, Anthony Ciro, Neil Geoghagen, Peter Kelly, David Shum, Michael Wyler, and Geoffrey Barger for help, advice and fruitful collaborations. Last but not least, I thank Mr Jason Chan for help with typing part of this chapter, and Dr Patrick Morcillo for critical reading.

References

Amarzguioui, M., Holen, T., Babaie, R. and Prydz, H. (2003) Tolerance for mutations and chemical modifications in a siRNA. *Nucleic Acids Res* **31**: 589–595.

Behlke, M.A. (2006) Progress towards *in vivo* use of siRNAs. *Mol Ther* **13**: 644–670.

Birmingham, A., Anderson, E.M., Reynolds, A., *et al*. (2006) 3′-UTR seed matches, but not overall identity, are associated with RNAi off-targets. *Nat Methods* **3**: 199.

Brummelkamp, T.R., Bernards, R. and Agami, R. (2002) A system for stable expression of short interfering RNAs in mammalian cells. *Science* **296**: 550–553.

Caplen, N.J., Parrish, S., Imani, F., Fire, A. and Morgan, R.A. (2001) Specific inhibition of gene expression by small double-stranded RNAs in invertebrate and vertebrate systems *PNAS* **98**: 9742.

Chi, J.-T., Chang, H.Y., Wang, N.N., Chang, D.S., Dunphy, N. and Brown, P.O. (2003) Genome-wide view of gene silencing by small interfering RNAs. *PNAS* **100**: 6343.

Chiu, Y.L. and Rana, T.M. (2002) RNAi in human cells: basic structural and functional features of small interfering RNA. *Mol Cell* **10**: 549–561.

Couzin, J. (2002) Small RNAs make big splash. *Science* **298**: 2296–2297.

Cullen, B.R. (2006) Enhancing and confirming the specificity of RNAi experiments. *Nat Methods* **3**: 677–681.

Czauderna, F., Fechtner, M., Aygun, H., Arnold, W., Klippel, A., Giese, K. and Kaufmann, J. (2003) Functional studies of the PI(3)-kinase signaling pathway employing synthetic and expressed siRNA. *Nucleic Acids Res* **31**: 670–682.

Efimov, V.A., Chakhmakhcheva, O.G. and Wickstrom, E. (2005) Synthesis and application of negatively charged PNA analogues. *Nucleosides, Nucleotides, Nucleic Acids* **24**: 1853–1874.

Elbashir, S.M., Haborth, J., Lendeckel, W., Yalcin, A., Weber, K. and Tuschl, T. (2001a) Duplexes of 21-nucleotide RNAs mediate RNA interference in cultured mammalian cells. *Nature* **411**: 494–498.

Elbashir, S.M., Lendeckel, W. and Tuschl, T. (2001b) RNA interference is mediated by 21- and 22-nucelotide RNAs. *Genes Dev* **15**: 188–200.

Elbashir, S.M., Martinez, J., Patkaniowska, A., Lendeckel, W. and Tuschl, T. (2001c) Functional anatomy of siRNAs for mediating efficient RNAi in *Drosophila melanogaster* embryo lysate. *EMBO J* **20**: 6877–6888.

Fire, A., Xu, S., Montgomery, M.K., Kostas, S.A., Driver, S.E., Mello, C.C. (1998) Potent and specific genetic interference by double-stranded RNA in *Caenorhabditis elegans*. *Nature* **391**: 806–811.

Hammond, S.M., Bernstein, E., Beach, D. and Hannon, G.J. (2000) An RNA-directed nuclease mediates posttranscriptional gene silencing in *Drosophila* cells. *Nature* **404**: 293–296.

Hannon, G.J. (2002) RNA interference. *Nature* **418**: 244–251.

Hogrefe, R., Lebedev, A., Zon, G., Pirollo, K., Rait, A., Zhou, Q., Yu, W. and Chang, E. (2006) Chemically modified short interfering hybrids (siHYBRIDS): nanoimmunoliposome delivery in vitro and in vivo for RNAi of HER-2. *Nucleosides, Nucleotides, Nucleic Acids* **25**: 889–907.

Holen, T., Amarzguioui, M., Babaie, E. and Prydz, H. (2003) Similar behavior of single-strand and double-strand siRNAs suggests they act through a common RNAi pathway. *Nucleic Acids Res* **31**: 2401–2407.

Jagla, B., Aulner, N., Kelly, P.D., *et al.* (2005) Sequence characteristics of functional siRNAs. *RNA* **11**: 864–872.

Jakson, A.L., Bartz, S.R., Scheller, J., Kobayashi, S.V., Burchard, J., Mao, M., Li, B., Cavet, G. and Linsley, P.S. (2003) Expression profiling reveals off-target gene regulation by RNAi. *Nat Biotechnol* **21**: 635.

Jorgensen, R. (1990) Altered gene expression in plants due to trans interactions between homologous genes. *Trends Biotechnol* **8**: 340–344.

Lee, N.S., Dohjima, T., Bauer, G., Li, H., Li, M.-J., Ehsani, A., Salvaterra, P. and Rossi, J. (2002) Expression of small interfering RNAs targeted against HIV-1 rev transcripts in human cells. *Nat Biotechnol* **20**: 500–505.

Martinez, J., Patkaniowska, A., Urlaub, H., Luhrmann, R. and Tuschl, T. (2002) Single-stranded antisense siRNAs guide target RNA cleavage in RNAi. *Cell* **110**: 563–574.

McManus, M.T. and Sharp, P.A. (2002) Gene silencing in mammals by small interfering RNAs. *Nat Rev Genet* **3**: 737–747.

McManus, M.T., Haines, B.B., Dillon, C.P., Whitehurst, C.E., van Parijs, L., Chen, J. and Sharp, P.A. (2002) Small interfering RNA-mediated gene silencing in T lymphocytes. *J Immunol* **169**: 5754–5760.

Moffat, J. and Sabatini, D.M. (2006) Building mammalian signaling pathways with RNAi screens. *Nat Rev Mol Cell Biol* **7**: 177–187.

Muench, U. and Pfleiderer, W. (2003) Oligoribonucleotide synthesis with the (2-cyano-1-phenylethoxy)carbonyl(2x1peoc) group for the 5'hydroxy protection. *Helv Chim Acta* **86**: 2546–2565.

Novina, C.D., Murray, M.F., Dykxhoorn, D., *et al.* (2002) siRNA-directed inhibition of HIV-1 infection. *Nat Med* **8**: 681–686.

Nykaneri, A., Haley, B. and Zamore, P.D. (2001) ATP requirements and small interfering RNA structure in the RNA interference pathway. *Cell* **107**: 309–321.

Ogilvie, K.K., Sadana, K.L., Thompson, A.E., Quillian, M.A. and Westmore, J.B. (1974) *Tetrahedron Lett* **15**: 2861–2863.

Ohgi, T., Masutomi, Y., Ishiyama, K., Kitagawa, H., Shiba, Y. and Yano, J. (2005) A new RNA synthetic method with a 2'-*O*-(2-cyanoethoxymethyl) protecting group. *Org Lett* **7**: 3477–3480.

Paddison, P.J., Caudy, A.A., Bernstein, E., Hannon, G.J. and Conklin, D.S. (2002)

Short hairpin RNAs (shRNA) induce sequence-specific silencing in mammalian cells. *Genes Dev* **16**: 948–958.

Pitsch, S., Weiss, P.A., Jenny, L., Stutz, A. and Wu, X. (2001) Reliable chemical synthesis of oligoribonucleotides (RNA) with 2-O-[(Triisopropylsilyl) oxy]methyl(2-O-tom)-protected phosphoramidites. *Helv Chim Acta* **84**: 3773–3795.

Pon, R.T., Yu, S., Prabhavalkar, T., Mishra, T., Kulkarni, B. and Sanghvi, Y.S. (2005) Large-scale synthesis of 'Cpep' RNA monomers and their application in automated RNA synthesis. *Nucleosides, Nucleotides, Nucleic Acids* **24**: 777–781.

Porcher, S. and Pitsch, S. (2005) Synthesis of 2'-O-[(triisopropylsilyl)oxy]methyl (TOM)-protected ribonucleoside phosphoramidites containing various nucleobase analogs. *Helv Chim Acta* **88**: 2683–2704.

Qin, X.F., An, D.S., Chen, I.S. and Baltimore, D. (2003) Inhibiting HIV-1 infection in human T cells by lentiviral-mediated delivery of small interfering RNA against CCR5. *PNAS USA* **100**: 183–188.

Reese, C.B. (2005) Oligo- and poly-nucleotides: 50 years of chemical synthesis. *Org Biomol Chem* **3**: 3851–3868, and the references cited therein.

Romano, N. and Macino, G. (1992) Quelling: transient inactivation of gene expression in *Neurospora crassa* by transformation with homologous sequences. *Mol Microbiol* **6**: 3343.

Sachse, C. and Echeverri, C.J. (2004) Oncology studies using siRNA libraries: the dawn of RNAi-based genomics. *Oncogene* **23**: 8384–8391.

Sandy, P., Ventura, A. and Jacks, T. (2005) Mammalian RNAi: a practical guide. *Biotechniques* **39**: 215–224.

Scaringe, S.A., Wincott, F.E. and Caruthers, M.H. (1998) Novel RNA synthesis method using 5'-O-silyl-2'-O-orthoester protecting groups. *J Am Chem Soc* **120**: 11820–11821.

Scherer, L. and Rossi, J.J. (2004) Therapeutic applications of RNA interference: recent advances in siRNA design. *Adv Genet* **5**: 1–21.

Scherer, L.J. and Rossi, J.J. (2006). Recent applications of RNAi in mammalian systems. In: *Peptide Nucleic Acids, Morpholinos, and Related Antisense Biomolecules* (eds C. Jensen and M. During). Landes Bioscience/Springer, Georgetown, Texas.

Schwarz, D.S., Hutvagner, G., Haley, B., Zamore, P.D. (2002) Evidence that siRNAs function as guides, not primers in the *Drosophila* and human RNAi pathways. *Mol Cell* **10**: 537–548.

Semizarov, D., Frost, L., Sarthy, A., Kroeger, P., Halbert, D.N., Fesik, S.W. (2003) Specificity of short interfering RNA determined through gene expression signature *PNAS* **100**: 6347.

Sharp, P.A. (2001) RNA interference – 2001. *Genes Dev* **15**: 485–490.

Smith, C. (2006) Sharpening the tools of RNA interference. *Nat Methods* **3**: 475–484.

Spagnou, S., Miller, A.D. and Keller, M. (2004) Lipidic carriers of siRNA: differences in the formulation, cellular uptake, and delivery with plasmid DNA. *Biochemistry* **43**: 13348–13356.

Tuschl, T. (2002) Expanding small RNA interference. *Nat Biotechnol* **20**: 446–448.

Utku, Y., Dehan, E., Ouerfelli, O., Piano, F., Zuckermann, R.N., Pagano, M. and Kirshenbaum, K. (2006) A peptidomimetic siRNA transfection reagent for highly effective gene silencing. *Mol BioSyst* **2**: 312–317.

Wilson, J.A., Jayasena, S., Khvorova, A., *et al.* (2003) RNA interference blocks gene expression and RNA synthesis from hepatitis C replicons propagated in human liver cells. *PNAS* **100**: 2793–2788.

Zamore, P.D. (2001) RNA interference: listening to the sound of silence. *Nat Struct Biol* **8**: 6047–6062.

Zamore, P.D. and Aronin, N. (2003) siRNAs knock down hepatitis. *Nat Med* **9**: 266–267.

Zamore, P.D., Tuschl, T., Sharp, P.A. and Bartel, D.P. (2000) RNAi: double-stranded RNA directs the ATP-dependent cleavage of mRNA at 21 to 23 nucleotide intervals. *Cell* **101**: 25–33.

Validation of RNAi

4

Nathalie Aulner and Bernd Jagla

4.1 Introduction

In recent years, RNAi has developed into a leading technique to assess gene function in eukaryotic cells (Mello and Conte, 2004). RNAi (also called post-transcriptional gene silencing) is a process in which a dsRNA triggers the degradation of a homologous mRNA. A long dsRNA is cleaved by the dsRNA processing enzyme Dicer into small 21–23mers, referred to as siRNA, which are incorporated into the RISC and unwound. When loaded with a single-strand siRNA, RISC* binds to a complementary sequence on an mRNA molecule and cleaves it between nucleotides 10 and 11 relative to the siRNA (Elbashir *et al.*, 2001b; Yuan *et al.*, 2005). This initiates the degradation of the target mRNA and, therefore, inhibits further gene expression. Mammalian cells have a cellular defense mechanism that, in the presence of dsRNAs (longer than 30 base pairs), provokes a global unspecific repression of gene expression (Sledz and Williams, 2004; Stark *et al.*, 1998). The discovery that small 21mer siRNA, in contrast to longer dsRNA, elicits a very limited unspecific response (Elbashir *et al.*, 2001a) allowed the use of the technology as a tool to assess gene function in mammalian cells. Because of its efficiency and high specificity, RNAi has revolutionized genomics and drug discovery. It has become the technique of choice to perform reverse genetics in organisms where previously genetic manipulation was difficult if not impossible. RNAi is easily scalable to study all genome functions and has proven useful for many applications, including functional annotation of genome data and *in vivo* target validation. Finally, therapeutic applications of RNAi are currently being studied intensively because of their potential for the development of gene-specific medicine (Huppi *et al.*, 2005; Mittal, 2004). To allow the successful delivery of the RNA duplexes into mammalian cell lines, different strategies have been developed over the last few years, including chemical synthesis (Elbashir *et al.*, 2002), *in vitro* transcription (Donze and Picard, 2002), or vector-based delivery (Miyagishi *et al.*, 2004).

siRNAs have to be highly efficient and as specific as possible to be used with confidence. Many algorithms are now available for the rational design of siRNA molecules *in silico* (see Chapter 2 and references therein), giving the researcher a higher chance to perform a successful knockdown.

It is, however, necessary to use this technology with caution, as many studies have shown some limitations such as efficient delivery and knockdown or control of the potential secondary effects (off-target and interferon response) (Huppi *et al.*, 2005). The validity of the probe used has, therefore, to be confirmed experimentally. In this chapter, we review the different delivery systems of RNAi effectors into mammalian cells, as well as the different ways to assess their validity, that is, efficacy and specificity, and the controls to perform and precautions to take to minimize any secondary effects, including knockdown of unwanted genes (off-target effect) and induction of a cell defense mechanism (interferon response).

4.2 siRNA delivery

In order to achieve a successful knockdown using RNAi, the delivery system for mammalian cells should be chosen carefully to allow the right amount of duplexes to enter the cells. Low transfection efficiency and low cell viability are in fact some of the most frequent causes of unsuccessful gene silencing experiments. Efficient delivery of RNAi probes can be achieved by several methods, including (i) direct transfection of siRNAs molecules or (ii) introduction of short hairpin RNA (shRNA) expressing plasmids. Additionally, when conducting RNAi screens, one should consider a delivery system suitable for high throughput (iii).

4.2.1 siRNA transfection

Like nucleic acids in general, small interfering RNAs are polyanions. Their unassisted permeation through the lipid bilayer of mammalian cells is, therefore, negligible. The uptake of siRNA by mammalian cultured cells appears to be fundamentally different than the commonly used DNA transfection (Spagnou *et al.*, 2004). Many approaches including chemical, biological, and physical systems have been successfully used to transiently deliver the siRNA molecules, even in more difficult systems such as primary cells. The choice of delivery system is therefore highly dependent on the targeted cell type. Lipid-based transfection, membrane-permeant peptides (MPPs), supramolecular nanocarrier, and electroporation are the most commonly used delivery systems. Lipid-based transfection is the method of choice for siRNA delivery into immortalized cultured cells. Many different formulations, including cationic, liposomal, and polyamide-based agents are commercially available. Despite their commercial availability, these reagents still must be tested for potential side-effects that can be caused, for example, by cell-type-specific factors. Optimizing the transfection conditions is the key to successful gene silencing experiments. The parameters include cell culture conditions, choice and amount of transfection reagent, transfection timing, and siRNA quantity and quality. It is important to follow the guidelines provided for each transfection technique. The use of cationic liposomes for transfection is highly efficient but is unfortunately not usable in a number of cell types including dendritic or endothelial cells. To circumvent this lack of efficiency, the use of MPPs can be considered. MPPs are short emphiphatic peptides that can translocate through lipid bilayers in an energy-dependent manner (Muratovska and Eccles, 2004). The delivery of the siRNA through this process is achieved by conjugating the MPP to a thiol group that has been added at the 5′ end of one of the strands. Other methods including the use of supramolecular nanocarriers (Itaka *et al.*, 2004; Schiffelers *et al.*, 2004), lipid and steroid conjugates (Lorenz *et al.*, 2004) have also been reported. It has been shown that the detachment of cells and co-incubation with the transfection complexes while in suspension increases the silencing efficiency for cell types that are refractory to typical lipid-based transfection (Amarzguioui, 2004). Electroporation is another efficient alternative to chemical transfection for primary cells, cells growing in suspension, and cell types that are difficult to transfect. Electroporation consists of the application of a brief electrical field pulse to induce transient cell membrane permeability via the formation of microscopic electropores. The electrical field pulses have to be optimized to allow the cells to recover and enough siRNA to transfect the cells.

4.2.2 Introduction of shRNAs into mammalian cells

Effective RNAi can not only be achieved by transfection of chemically synthesized siRNAs, but also by introducing an shRNA expressing plasmid. In this case, a DNA vector contains appropriate promoter and termination sequences that flank the guide and passenger strands, which are in turn linked by a small number of nucleotides (linker region). The sequence is transcribed by the intracellular machinery into an shRNA; the shRNA is then processed by Dicer before entering RISC and causing RNAi (Donze and Picard, 2002). This method has several advantages over direct siRNA transfection, including a more pronounced and long-lived RNAi effect and the possibility to control the timing of the shRNA expression with inducible promoters (Huppi *et al.*, 2005). Initially, shRNA expression vectors used RNA pol III promoter and termination sequences (Donze and Picard, 2002). More recently, expression systems have been developed utilizing flanking sequences enabling RNA pol II expression, based on better understanding of the expression of intracellular micro RNAs (miRNAs) (Paddison *et al.*, 2004; Shinagawa and Ishii, 2003). This has drastically increased its potency, allowing for the cell type-specific and temporal regulation of the shRNA expression. The latter enables control of the initiation of the RNAi effect, the study of phenotypic changes during recovery, and, more importantly, the study of loss of function of genes required for cell viability and proliferation. A major drawback is that plasmid transfection is much less efficient than direct siRNA delivery (Spagnou *et al.*, 2004).

Chemical, biological, and physical reagents, among others, are commonly used to transfect plasmid DNA into mammalian cells. Many of them are commercially available. As for the direct transfection of siRNAs, their choice is highly dependent on the cell type. The transfection conditions (quantity of plasmid DNA, ratio to carrier and other considerations) have to be established in accordance with the manufacturer's guidelines. Interestingly, the plasmid-based system can be coupled with viral delivery systems. For this purpose, a vector containing appropriate viral packaging signals and regulatory elements is used to package the shRNA sequence into infectious virions. These viral particles transduce a broader spectrum of cell lines and can overcome many issues of standard transfection methods. Adenovirus and a number of retroviruses such as lentivirus and murine stem cell virus (MSCV) are a few of the most commonly used viral delivery systems (Brummelkamp *et al.*, 2002; Rubinson *et al.*, 2003; Shen *et al.*, 2003). Adenoviruses utilize receptor-mediated infection and do not integrate into the genome, while MSCV cannot integrate into non-dividing cell lines such as neurons. The lentiviral system on the other hand has the ability to integrate into the genome and therefore to create stable gene silencing. In addition, it does not require a mitotic event for integration into the genome and it can be used in both dividing and non-dividing cell lines.

4.2.3 RNAi screening delivery systems

Because of its ease of use, RNAi has been adopted in high throughput screening of gene function analysis (Carpenter and Sabatini, 2004), pathway analysis and drug target validation (Ovcharenko et al., 2005). For this purpose, numerous si/shRNA libraries have been developed (Ren et al., 2006; Sachse and Echeverri, 2004; Silva et al., 2005; Xin et al., 2004). There are important considerations for a successful siRNA library screen. Transfection efficiency, reproducibility, and cell viability relate directly to the transfection method. New technologies have been developed to meet the criteria necessary for high-throughput screening using RNAi. Among these methods, reverse transfection and 96-well plate electroporation are the most commonly used techniques to date. Reverse transfection can be used for most immortalized adherent cell lines (Amarzguioui, 2004; Ovcharenko et al., 2005), whereas electroporation is suitable for most primary cells and cells grown in suspension. The latter necessitates, however, the purchase of costly specific instrumentation. Recently, the use of microarray-based transfection allowing the parallel delivery of siRNA libraries has been implemented in several laboratories (Mousses et al., 2003; Wheeler et al., 2005). In this case, siRNA transfection complexes are spotted on a glass slide and cells are plated on the slide. Only the cells in close proximity to the complexes will be transfected. This methodology can also be used for transfecting vector-based shRNA (Mittal, 2004). Direct delivery of shRNAs into mammalian cells via integrin-receptor mediated bacterial invasion has been reported (Zhao et al., 2005). The shRNA is in this case produced directly in an 'invasive' E. coli strain. The strain has been engineered by knocking-in the Yersinia pseudoturberculosis gene coding for invasin (invasin binds to the integrin receptor of mammalian cells) (Grillot-Courvalin et al., 1998). Under the experimental conditions used by Zhao and colleagues, bacterial invasion does not elicit more interferon response than the direct transfection of siRNA at the right concentration. This method is particularly convenient for screening large shRNA libraries, as it does not require any plasmid purification and is therefore very cost-effective and fast.

For all the above-mentioned delivery techniques, it is important to consider the amount of si/shRNA that will be delivered to the cells and the timing of the experiment. Different degrees of silencing are produced by different doses of silencing effectors as well as by the time point at which the knockdown is assessed post-transfection (Raab and Stephanopoulos, 2004). Lowering the siRNA dose is also likely to reduce secondary effects such as off-target gene silencing (as discussed in a following paragraph) and potential toxicity.

Several methods can be used to assess the quality of the transfection, that is, the validation of uptake of interfering molecules and to screen for efficient knockdown among the cell population. These are dependent on the technology chosen to induce RNAi. When using siRNAs, the most cost-effective and commonly used method is parallel transfection of a known efficient siRNA against a target that is easily measurable and functionally unrelated. This gives a rough estimate of the transfection efficiency for the screened siRNA and should be considered a control experiment. The use of fluorescently end-labeled siRNA (Chiu *et al.*, 2004) or cotransfection with reporter plasmid has also been described. This allows selecting a subpopulation of cells by fluorescence and/or antibiotic resistance (Kumar *et al.*, 2003). However, siRNA modifications are quite expensive and the level of intracellular fluorescent molecules needed for accurate measurement is most of the time several dimensions higher than what is needed for efficient knockdown. Cotransfection with a reporter plasmid might not efficiently correlate with siRNA uptakes, as plasmid DNA and siRNA transfection are two different phenomena. Interestingly, Chen and colleagues (Chen *et al.*, 2005) have shown that the use of semiconductor quantum dots (QD) circumvent some of the above-mentioned problems. QD are bright, photostable fluorescent nanocrystals that are non-toxic and brighter than conventional fluorophores, as well as cost effective. Because they are compatible with a variety of transfection methods, the degree of silencing can easily be correlated with the cellular fluorescence. This methodology allows collecting a uniformly silenced cell subpopulation by fluorescence-activated cell sorting (FACS) (Chen *et al.*, 2005). On the other hand, the effective delivery of shRNA can be easily measured by the addition of a fluorescent reporter to the vector carrying the hairpin sequence (Malik *et al.*, 2006).

Once the transfection method has been chosen and the delivery conditions have been established, one should consider using single or multiple transfections of siRNA and assaying at different time points to take into account the half-life of the targeted protein and the desired degree of silencing. It has, in fact, been shown that for long-lived proteins, multiple transfections over several days or longer incubation times were necessary to achieve a measurable reduction in expression (Choi *et al.*, 2005).

4.3 Silencing efficacy (potency)

A good RNAi experiment is not only dependent on the delivery system but also on the ability of the siRNA/shRNA to effectively knock down the targeted gene. The technology should be used with caution, especially when working on a transcript with high rates of polymorphism; it is probable that an RNAi probe will work in one context (cell line) and not as efficiently in another. Many laboratories and commercial companies have developed *in silico* RNAi probe prediction tools (see Chapter 2 and Jagla *et al.*, 2005), usually based on sequence-specific and thermodynamic parameters. These prediction algorithms are unfortunately not foolproof and researchers are left with the task to confirm that the siRNA used is indeed knocking down the intended target. For this purpose, many laboratories use conventional methods or have developed methodologies to measure the levels of either the mRNA or the corresponding protein.

The knockdown efficiency can be controlled either by measuring the levels of (i) the endogenous target mRNA; (ii) the corresponding protein levels; or (iii) by using a reporter plasmid carrying the targeted sequence fused to a reporter gene. Each of these techniques has advantages and disadvantages as describe below. The choice of the validation method is dependent on the type of study being performed. Finally, a validation scheme for an si/shRNA screen can only be achieved with a technology suitable for high-throughput.

4.3.1 Detection of mRNA levels

The primary target of the RNAi effector is the mRNA; therefore, the most obvious control should be the quantification of the target mRNA in the cell after si/shRNA treatment. Detecting the mRNA levels could, however, yield to erroneous results, as little is known about the kinetics and mechanism of the targeted mRNA degradation after the initial endonucleolytic cleavage guided by the siRNA. It has been shown by Northern Blot analysis that some intermediary products can accumulate for a certain amount of time (Holen *et al.*, 2002). The levels of the endogenous target mRNA can be measured by Northern Blot analysis, ribonuclease (RNase) protection, reverse transcriptase (RT)-PCR, quantitative RT-PCR (qRT-PCR), branched DNA (bDNA), or microarrays. In all cases, total RNA or mRNAs are prepared at the desired time point after si/shRNA transfection (typically 24–48 h) and assayed with the chosen methodology. It seems that at longer time points (24 h and up), the amount of intermediary products becomes insignificant (Holen *et al.*, 2002).

Northern Blot analysis is an informative procedure to confirm that there are no more partially degraded transcripts. After being separated by electrophoresis, the cellular RNA is transferred to a membrane; then the level of targeted mRNA and the level of a well-chosen internal control mRNA (e.g. coding for a house keeping gene and not functionally related to the target) are detected with specific nucleic acid probes. The level of remaining mRNA is then normalized to the internal control.

RNase protection assays (RPA) can also be very informative but are relatively tedious to perform for laboratories not familiar with molecular biology or working with RNA. The cellular mRNAs are incubated with labeled antisense probes specific to the targeted mRNAs and a well-chosen internal control mRNA. Unhybridized RNAs are then removed by RNase I digestion. The protected fragment is analyzed by electrophoresis. This technique is more sensitive than Northern Blot analysis.

RT-PCR can also be used to quantify the remaining amount of the targeted mRNA. For this purpose, the whole cellular mRNA pool is reverse transcribed into complementary DNA (cDNA) and the cDNA corresponding to the targeted mRNA is amplified by PCR using specific probes. The quantification should be validated by normalization to an internal control. The choice of oligo-dT or random hexamers for the reverse transcription step should be made according to the location of the cutting site on the mRNA. The primer pairs for the PCR step should ideally surround the targeted site to avoid any misinterpretation of the results, especially when probing at shorter time points (Hahn *et al.*, 2004). RT-PCR is an 'end-point' method and is not very sensitive: small levels of variation may not be detectable. qRT-PCR, on the other hand, is a much more sensitive method but requires sequence-specific primers and a proprietary probe set, in addition to having access to appropriate hardware. Any quantitative PCR procedure should be carefully optimized for each mRNA, especially for low-abundance transcripts for which silencing might only represent a small change in the PCR cycle threshold. qRT-PCR provides several advantages for monitoring target mRNA levels. It is at least three orders of magnitude more sensitive than Northern Blot analysis and results can be obtained much more quickly. Additionally, the method provides quantitative results when carefully optimized.

The bDNA technology is similar in principle to an enzyme-linked immunosorbent assay (ELISA), using signal amplification to detect the target mRNA by measuring the signal generated by many branched, labeled DNA probes. The technology is derived from a technique used to measure viral load (Zhang *et al.*, 2005). Because of the signal amplification step, this technique is relatively sensitive but requires proprietary probe sets.

Microarray technology has been successfully used to validate RNAi (Mousses *et al.*, 2003). The principal advantage of this method is the ability to monitor global changes in transcription levels of the whole genome after RNAi and therefore enable the quantification of potential down-regulation of unwanted mRNA (off-target effects; described below).

The three latter methods (RT-PCR, bDNA, and microarray analysis) can easily be used as validation for RNAi screens, as they are suitable for parallel experimentations.

4.3.2 Detection of protein levels

The ultimate goal of an RNAi experiment is to decrease the amount of a specific protein in a cell and to analyze its biological impact. For this purpose, it is important to measure the amount of remaining targeted protein after RNAi treatment. The most common assays used are Western Blot and immunofluorescence. The major drawback of the two methods is the availability of antibodies targeting the protein of interest. As is the case for the Northern Blot analysis, Western Blot and immunofluorescence experiments necessitate the use of a well-chosen internal control and are only semi-quantitative. For most of the cases, mRNA and protein detection results correlate (Mittal, 2004) but, in some cases, probing the target protein will give useful additional information. Some proteins have a long half-life and even a very good mRNA knockdown might not exert the desired decrease of the targeted protein. In this case, it might be useful to probe at later time points, to probe multiple siRNA transfections over time, or to use shRNA methodology, which has been shown to last longer.

Probing for protein levels introduces challenges when conducting RNAi screens or when measuring the effect on all proteins. Methods allowing the quantification of whole cellular proteins have been recently developed and have been used for RNAi effect measurements. They use isotopic labeled peptides in conjunction with mass spectrometry (LC-MS) (Yao *et al.*, 2001).

4.3.3 Detection of knockdown efficiency using a reporter system (surrogate assays)

Many additional methodologies have been developed to validate the efficiency of a given siRNA. Most of them use a plasmid that ectopically expresses the targeted transcript or a portion of it fused to a reporter gene. This allows the rapid and cost effective test of as many si/shRNAs as one can handle. The basic idea behind this technology is using a reporter protein that is either directly visible (fluorescent protein, like GFP) or easy to measure (enzymatic assay, e.g. luciferase) (Malik *et al.*, 2006; Smart *et al.*, 2005). Many flavors of such systems have been developed and are commercially available. They allow quick and easy subcloning of either the whole cDNA or a portion of interest (containing the cutting site) into a reporter plasmid. The use of the whole targeted cDNA is dependent on the availability of the full length clone. However, in many cases, an additional sequencing step is necessary to validate its sequence. Du and coworkers have developed a methodology that circumvents this problem. They engineered a plasmid to enable the fast and easy cloning of a short oligonucleotide (19–38 base pairs) harboring the cutting site just after the start codon of the luciferase gene (Du *et al.*, 2004). In another study, Panstruga *et al.* co-expressed two fluorescent proteins with different spectra, one as a translational fusion with the target cDNA and the other as a transfection marker (Panstruga *et al.*, 2003).

The use of such surrogate validation systems is dependent on many parameters. The dynamic range of regulation should be large enough to be able to discriminate between siRNAs with different knockdown efficiencies. The constructs should allow the cloning of a large enough targeted sequence to take into account potential secondary structure effects. Finally, the cell type in which the test is performed should allow straightforward, high-efficiency transfection of plasmid DNA.

One major drawback of these systems is that several factors, such as secondary structure of the targeted mRNA and bound proteins might differ from the surrogate target. In this case, reporter-based strategies would give an erroneous answer.

When performing an RNAi screen, the si/shRNAs library should be validated. This is to avoid as many false positive as possible in the planned screen. For this purpose, Kumar and his colleagues have taken advantage of RNAi microarrays to verify siRNAs in a highly parallel assay (Kumar *et al.*, 2003).

When using one of these surrogate assays to validate an siRNA, follow-up studies testing the endogenous target should be performed in order to test the silencing efficacy in an endogenous context because the level of expression of the transgene might not reflect the actual transcription levels.

4.4 Silencing validation

The ultimate way to confirm the efficacy and specificity of a knockdown experiment, as has been used for years in classical genetic loss-of-function studies, is a rescue experiment (Sarov and Stewart, 2005). To achieve this goal, the reintroduced gene has to be resistant to the RNAi effect and its expression levels should be as close as possible to physiological levels. To generate RNAi-resistant constructs, one either generates point mutations in the cDNA, or in cases where the siRNA targets are in the untranslated region of a gene, this region can be omitted in the rescue plasmid (Lassus *et al.*, 2002). These approaches have certain disadvantages: introducing silent mutations can be time consuming and laborious, the expression levels, usually driven by a viral promoter, are non-physiological, and the potential alternative splicing of the target gene is lost. Kittler and coworkers have developed a system based on bacterial artificial chromosomes (BACs) carrying closely related orthologous genes that circumvent these potential drawbacks (Kittler *et al.*, 2005). BACs carry long portions of the genome allowing physiological expression of all potential alternative splice variants. When using a genome from a closely related species, it is likely that the sequence already has the necessary sequence differences that will prohibit the siRNA from targeting it.

4.5 siRNA specificity

One major source of concern of RNAi is the target specificity and potential off-target effects, specifically targeting of other unintended mRNAs in the cells. The risks are many-fold: off-target effects due to partial complementarity, miRNA effects (partial complementarity leading to translational inhibition), and activation of the innate immune response. The latter is, for the most part, sequence independent and will be discussed in a following paragraph. The unintended off-target silencing is more widely spread as previously thought and seems to occur, at least partially, in a manner reminiscent of target silencing by miRNA (translational silencing) (Jackson *et al.*, 2006a, 2006b). The potential effect of an si/shRNA on other mRNAs than the intended target can be shown by microarray analysis. This technique allows measuring whole genome expression levels. It has been shown that mRNAs with as few as 11 consecutive nucleotides matching the siRNA sequence can be down-regulated (Jackson *et al.*, 2003). Yet another possible artifact can be produced when the intended passenger strand enters RISC and acts as a guide strand to elicit silencing of a matching target mRNA. Unfortunately, many published microarray studies have shown contradictory results (Chi *et al.*, 2003; Jackson *et al.*, 2003; Persengiev *et al.*, 2004; Semizarov *et al.*, 2004). Moreover, miRNA generated off-target effects are difficult to assess as miRNAs exert their function at the translational level (Scacheri *et al.*, 2004). It is therefore highly recommended to design the sequence of the RNAi probe very carefully, especially when microarrays are not available. More specifically, to avoid the entrance of the passenger strand into the RISC complex, one should control the thermodynamic asymmetry of the si/shRNA molecule (see Chapter 3).

Recently, a couple of studies have given new insights into this matter. Jackson and colleagues have detected a strong correlation between off-target silencing and partial complementarity at the 5'-end of the guide strand (Jackson *et al.*, 2006a). This portion of the siRNA acts similarly to the seed sequence of miRNA (Lin *et al.*, 2005) and seems to be a primary determinant of the off-target effect. This result is independent of both the concentration and the technology used to induce RNAi as the same off-target was observed by using shRNA. In addition, it occurs in multiple cell types (Jackson *et al.*, 2006a). In an accompanying study, Jackson and colleagues (Jackson *et al.*, 2006a) described a position-specific chemical modification scheme to remove this partial complementarity without affecting the perfect match target. Introducing 2'-*O*-methyl modifications to specific positions in the seed region can, indeed, reduce the number of off-target genes as well as the magnitude of their down-regulation in a manner superior to mismatches. In addition, it has been shown that some siRNAs induce global changes in cell viability in a target independent fashion (Fedorov *et al.*, 2006). This can be very troublesome in screening campaigns, as off-target effects could be responsible for a large portion of the hits. The induced reduction of cell viability is diminished when the concentration of siRNA is reduced. To resolve this issue, Fedorov and coworkers propose another chemical modification pattern to significantly reduce this off-target effect with minimal influence on the targeted genes (Fedorov *et al.*, 2006). To confirm that the intended target is silenced and not another mRNA, it is useful to confirm the phenotype observed by using different independent siRNAs targeting the same mRNA. In addition, the specificity should be checked by BLASTing the si/shRNA sequence against the genome of interest. Details on the *in silico* method to verify siRNA specificity are described in another chapter of this book (Chapter 2). Finally, it is important to note that some 'off-target' effects might represent genuine physiological knock-on effects of specific target knockdown in certain pathways/signaling cascades. This is why one should interpret results carefully; confirm data with multiple siRNA, scrambled siRNA and/or other techniques when available.

4.6 Minimizing cell defense mechanism (dsRNA interferon response)

Mammalian systems have an innate defense mechanism directed against dsRNA (Bagasra and Prilliman, 2004). Because the introduction of long dsRNA molecules induces a global, non-specific suppression of gene expression as well as the expression of interferon responding genes, it was thought at first that RNAi could not be used in mammalian systems. The discovery of Tuschl's laboratory (Elbashir *et al.*, 2001a), that 21–23mers siRNA can bypass this defense mechanism, has allowed the use of RNAi in mammalian cell systems. Unfortunately, it appears that even these small dsRNA molecules can, in certain cases, elicit an interferon response (Sledz *et al.*, 2003). Little is know about the mechanism, but recent studies have suggested a new understanding.

The two best characterized dsRNA-induced pathways are found in most mammalian cell types and elucidate their signaling cascade through the dsRNA-dependent kinase protein kinase R (PKR) and 2',5'-oligoadenylate synthase (OAS1). PKR autophosphorylates in response to dsRNA and subsequently phosphorylates eIF2 (eukaryotic initiation factor 2) leading to global translation inhibition (Saunders and Barber, 2003). The activation of 2',5'-oligoadenylate synthase leads to the formation of 2',5'-oligoadenylates which bind and activate RNaseL. RNaseL then cleaves both cellular and non-cellular RNA in a non-specific manner (Pebernard and Iggo, 2004).

In addition to inhibiting translation, PKR is also a signal transducer leading to the activation of interferon response genes through the nuclear factor κ B (NFκB), Janus kinase–signal transducers and activators of transcription (JAK-STAT), and interferon regulatory factor-3 (IRF-3) signaling cascades (Sledz *et al.*, 2003). The interferon system is a cellular defense mechanism against viral infection that, when sufficiently activated, can cause an arrest in protein synthesis and lead to cell death (Stark *et al.*, 1998). Activation of this system could therefore complicate the interpretation of some experiments, as well as pose major problems for the use of siRNAs as therapeutics (Mousses *et al.*, 2003).

Recently, several reports have shown that siRNAs can induce the expression of components of the interferon system in animal cells (Sledz *et al.*, 2003). Transfection of some DNA vectors that express shRNAs, which are processed into siRNAs in the cell, can also induce expression of classical interferon response genes such as OAS1 (Bridge *et al.*, 2003) or stimulate type 1 interferon, interleukin 8 (IL-8), and tumor necrosis factor-alpha (TNFα) production (Kariko *et al.*, 2004). Similarly, transfection of *in vitro* transcribed siRNAs into cultured mammalian cells resulted in the induction of Stat1, a transcription factor involved in mediating the induction of interferon-stimulated genes (Sledz *et al.*, 2003), or an increase in interferon-β levels (Kim *et al.*, 2004).

To confirm that a given si/shRNA does not elicit an interferon response, a couple of critical controls should be performed. If available, a microarray study will show if classical interferon response genes are up-regulated after siRNA treatment (Scacheri *et al.*, 2004). Performing a Northern Blot analysis against a couple of classical interferon response genes such as OAS1 (Pebernard and Iggo, 2004) will test for relevant proteins. Alternatively, measuring the amount of interferon α or β released in the media after RNAi treatment (Kim *et al.*, 2004) are good options. A couple of precautions can be taken to minimize these undesirable effects like lowering the concentration of the RNAi effector. This can be achieved by using chemically or enzymatically synthesized siRNA but is more challenging when using a vector-driven shRNA (the promoter driving the expression of the shRNA should be chosen carefully and allow the control of the expression).

4.7 Conclusion

One of the best ways to study and to understand the function of a gene product is to prevent its expression or activity by one of several loss-of-function approaches. RNAi has become one of the leading techniques to do so, as witnessed by the amount of recent publications using the technology. However, behind this apparent ease, lie many obstacles mostly due to the partial knowledge of the intracellular mechanisms being used to elicit the gene silencing. Recent progress towards understanding of the cellular mechanism behind RNAi has counterbalanced some initial limitations.

In June 2003, an editorial in *Nature and Cell Biology* (Editorial, 2003) has pinpointed the necessity to ask investigators using RNAi to disclose some of the following controls: (i) use of a control siRNA (either mismatch or scrambled); (ii) perform basic controls such a diminution of targeted mRNA or protein levels; (iii) titrate the siRNA to the lowest effective concentration to avoid side-effects (off-target or interferon response); (iv) multiply the siRNA targeting the gene of interest (two to three different siRNAs targeting the same mRNA showing the same phenotype can validate the experimental results); (v) perform a functional control (rescue by expression of the target gene in a form refractory to the interference); and (vi) multiply the controls (the results/observed phenotype should be controlled by any means available).

Following these guidelines and correctly planning and controlling the experiment should lead to successful gene knockdown even in the more refractory cell lines and help the researcher to elucidate the function of their genes of interest.

Acknowledgments

The RNAi literature being so prolific, we apologize to the many investigators whose work was not cited in this manuscript. We would like to thank Drs Matthew Beard, Udo Többen, and Allen Volchuk for helpful discussions and critical reading of the manuscript.

References

Amarzguioui, M. (2004) Improved siRNA-mediated silencing in refractory adherent cell lines by detachment and transfection in suspension. *Biotechniques* **36:** 766–768, 770.

Bagasra, O. and Prilliman, K.R (2004) RNA interference: the molecular immune system. *J Mol Histol* **35:** 545–553.

Bridge, A.J., Pebernard, S., Duraux, A., Nicoulaz, A.L. and Iggo, R. (2003) Induction of an interferon response by RNAi vectors in mammalian cells. *Nature Genet* **34:** 263–264.

Brummelkamp, T.R., Bernards, R. and Agami, R. (2002) A system for stable expression of short interfering RNAs in mammalian cells. *Science* **296:** 550–553.

Carpenter, A.E. and Sabatini, D.M. (2004) Systematic genome-wide screens of gene function. *Nature Rev Genet* **5:** 11–22.

Chen, A.A., Defus, A.M., Khetani, S.R. and Bhatia, S.N. (2005) Quantum dots to monitor RNAi delivery and improve gene silencing. *Nucleic Acids Res* **33:** e190.

Chi, J.T., Chang, H.Y., Wang, N.N., Chang, D.S., Dunphy, N. and Brown, P.O. (2003) Genomewide view of gene silencing by small interfering RNAs. *Proc Natl Acad Sci USA* **100:** 6343–6346.

Chiu, Y.L., Ali, A., Chu, C.Y., Cao, H. and Rana, T.M. (2004) Visualizing a correlation between siRNA localization, cellular uptake, and RNAi in living cells. *Chem Biol* **11:** 1165–1175.

Choi, I., Cho, B.R., Kim, D., *et al.* (2005) Choice of the adequate detection time for the accurate evaluation of the efficiency of siRNA-induced gene silencing. *J Biotechnol* **120:** 251–261.

Donze, O. and Picard, D. (2002) RNA interference in mammalian cells using siRNAs synthesized with T7 RNA polymerase. *Nucleic Acids Res* **30:** e46.

Du, Q., Thonberg, H., Zhang, H.Y., Wahlestedt, C. and Liang, Z. (2004) Validating siRNA using a reporter made from synthetic DNA oligonucleotides. *Biochem Biophys Res Commun* **325:** 243–249.

Editorial (2003) Whither RNAi? *Nat Cell Biol* **5:** 489–490.

Elbashir, S.M., Harborth, J., Lendeckel, W., Yalcin, A., Weber, K. and Tuschl, T. (2001a) Duplexes of 21-nucleotide RNAs mediate RNA interference in cultured mammalian cells. *Nature* **411:** 494–498.

Elbashir, S.M., Martinez, J., Patkaniowska, A., Lendeckel, W. and Tuschl, T. (2001b) Functional anatomy of siRNAs for mediating efficient RNAi in *Drosophila melanogaster* embryo lysate. *EMBO J* **20**: 6877–6888.

Elbashir, S.M., Harborth, J., Weber, K. and Tuschl, T. (2002) Analysis of gene function in somatic mammalian cells using small interfering RNAs. *Methods* **26**: 199–213.

Fedorov, Y., Anderson, E.M., Birmingham, A., Reynolds, A., Karpilow, J., Robinson, K., Leake, D., Marshall, W.S. and Khvorova, A. (2006) Off-target effects by siRNA can induce toxic phenotype. *RNA* **12**: 1188–1196.

Grillot-Courvalin, C., Goussard, S., Huetz, F., Ojcius, D.M. and Courvalin, P. (1998) Functional gene transfer from intracellular bacteria to mammalian cells. *Nat Biotechnol* **16**: 862–866.

Hahn, P., Schmidt, C., Weber, M., Kang, J. and Bielke, W. (2004) RNA interference: PCR strategies for the quantification of stable degradation-fragments derived from siRNA-targeted mRNAs. *Biomol Eng* **21**: 113–117.

Holen, T., Amarzguioui, M., Wiiger, M.T., Babaie, E. and Prydz, H. (2002) Positional effects of short interfering RNAs targeting the human coagulation trigger Tissue Factor. *Nucleic Acids Res* **30**: 1757–1766.

Huppi, K., Martin, S.E. and Caplen, N.J. (2005) Defining and assaying RNAi in mammalian cells. *Mol Cell* **17**: 1–10.

Itaka, K., Kanayama, N., Nishiyama, N., Jang, W.D., Yamasaki, Y., Nakamura, K., Kawaguchi, H., Kataoka, K. (2004) Supramolecular nanocarrier of siRNA from PEG-based block catiomer carrying diamine side chain with distinctive pKa directed to enhance intracellular gene silencing. *J Am Chem Soc* **126**: 13612–13613.

Jackson, A.L., Bartz, S.R., Schelter, J., Kobayashi, S.V., Burchard, J., Mao, M., Li, B., Cavet, G. and Linsley, P.S. (2003) Expression profiling reveals off-target gene regulation by RNAi. *Nature Biotechnol* **21**: 635–637.

Jackson, A.L., Burchard, J., Leake, D., *et al.* (2006a) Position-specific chemical modification of siRNAs reduces 'off-target' transcript silencing. *RNA* **12**: 1197–1205.

Jackson, A.L., Burchard, J., Schelter, J., Chau, B.N., Cleary, M., Lim, L. and Linsley, P.S. (2006b) Widespread siRNA 'off-target' transcript silencing mediated by seed region sequence complementarity. *RNA* **12**: 1179–1187.

Jagla, B., Aulner, N., Kelly, P.D., *et al.* (2005) Sequence characteristics of functional siRNAs. *RNA* **11**: 864–872.

Kariko, K., Bhuyan, P., Capodici, J. and Weissman, D. (2004) Small interfering RNAs mediate sequence-independent gene suppression and induce immune activation by signaling through toll-like receptor 3. *J Immunol* **172**: 6545–6549.

Kim, D.H., Longo, M., Han, Y., Lundberg, P., Cantin, E. and Rossi, J.J. (2004) Interferon induction by siRNAs and ssRNAs synthesized by phage polymerase. *Nat Biotechnol* **22**: 321–325.

Kittler, R., Pelletier, L., Ma, C., Poser, I., Fischer, S., Hyman, A.A. and Buchholz, F. (2005) RNA interference rescue by bacterial artificial chromosome transgenesis in mammalian tissue culture cells. *Proc Natl Acad Sci USA* **102**: 2396–2401.

Kumar, R., Conklin, D.S. and Mittal, V. (2003) High-throughput selection of effective RNAi probes for gene silencing. *Genome Res* **13**: 2333–2340.

Lassus, P., Rodriguez, J. and Lazebnik, Y. (2002) Confirming specificity of RNAi in mammalian cells. *Sci STKE* **2002**: PL13.

Lin, X., Ruan, X., Anderson, M.G., McDowell, J.A., Kroeger, P.E., Fesik, S.W. and Shen, Y. (2005) siRNA-mediated off-target gene silencing triggered by a 7 nt complementation. *Nucleic Acids Res* **33**: 4527–4535.

Lorenz, C., Hadwiger, P., John, M., Vornlocher, H.P., Unverzagt, C. (2004) Steroid and lipid conjugates of siRNAs to enhance cellular uptake and gene silencing in liver cells. *Bioorg Med Chem Lett* **14**: 4975–4977.

Malik, I., Garrido, M., Bahr, M., Kugler, S. and Michel, U. (2006) Comparison of test systems for RNA interference. *Biochem Biophys Res Commun* **341**: 245–253.

Mello, C.C. and Conte, D.J.R. (2004) Revealing the world of RNA interference. *Nature* **431**: 338–342.

Mittal, V. (2004) Improving the efficiency of RNA interference in mammals. *Nat Rev Genet* **5**: 355–365.

Miyagishi, M., Matsumoto, S. and Taira, K. (2004) Generation of an shRNAi expression library against the whole human transcripts. *Virus Res* **102**: 117–124.

Mousses, S., Caplen, N.J., Cornelison, R., *et al.* (2003) RNAi microarray analysis in cultured mammalian cells. *Genome Res* **13**: 2341–2347.

Muratovska, A. and Eccles, M.R. (2004) Conjugate for efficient delivery of short interfering RNA (siRNA) into mammalian cells. *FEBS Lett* **558**: 63–68.

Ovcharenko, D., Jarvis, R., Hunicke-Smith, S., Kelnar, K. and Brown, D. (2005) High-throughput RNAi screening *in vitro*: from cell lines to primary cells. *RNA* **11**: 985–993.

Paddison, P.J., Cleary, M., Silva, J.M., Chang, K., Sheth, N., Sachidanandam, R. and Hannon, G.J. (2004) Cloning of short hairpin RNAs for gene knockdown in mammalian cells. *Nat Methods* **1**: 163–7.

Panstruga, R., Kim, M.C., Cho, M.J. and Schulze-Lefert, P. (2003) Testing the efficiency of dsRNAi constructs in vivo: a transient expression assay based on two fluorescent proteins. *Mol Biol Rep* **30**: 135–140.

Pebernard, S. and Iggo, R.D. (2004) Determinants of interferon-stimulated gene induction by RNAi vectors. *Differentiation* **72**: 103–111.

Persengiev, S.P., Zhu, X. and Green, M.R. (2004) Nonspecific, concentration-dependent stimulation and repression of mammalian gene expression by small interfering RNAs (siRNAs). *RNA* **10**: 12–18.

Raab, R.M. and Stephanopoulos, G. (2004) Dynamics of gene silencing by RNA interference. *Biotechnol Bioeng* **88**: 121–132.

Ren, Y., Gong, W., Xu, Q., Zheng, X., Lin, D., Wang, Y. and Li, T. (2006) siRecords: an extensive database of mammalian siRNAs with efficacy ratings. *Bioinformatics* **22**: 1027–1028.

Rubinson, D.A., Dillon, C.P., Kwiatkowski, A.V., *et al.* (2003) A lentivirus-based system to functionally silence genes in primary mammalian cells, stem cells and transgenic mice by RNA interference. *Nat Genet* **33**: 401–406.

Sachse, M. and Echeverri, C.J. (2004) Oncology studies using siRNA libraries: the dawn of RNAi-based genomics. *Oncogene* **23**: 8384–8391.

Sarov, M. and Stewart, A.F. (2005) The best control for the specificity of RNAi. *Trends Biotechnol* **23**: 446–448.

Saunders, L.R. and Barber, G.N. (2003) The dsRNA binding protein family: critical roles, diverse cellular functions. *FASEB J* **17**: 961–983.

Scacheri, P.C., Rozenblatt-Rosen, O., Caplen, N.J., *et al.* (2004) Short interfering RNAs can induce unexpected and divergent changes in the levels of untargeted proteins in mammalian cells. *Proc Natl Acad Sci USA* **101**: 1892–1897.

Schiffelers, R.M., Ansari, A., Xu, J., *et al.* (2004) Cancer siRNA therapy by tumor selective delivery with ligand-targeted sterically stabilized nanoparticle. *Nucleic Acids Res* **32**: e149.

Semizarov, D., Kroeger, P. and Fesik, S. (2004) siRNA-mediated gene silencing: a global genome view. *Nucleic Acids Res* **32**: 3836–3845.

Shen, C., Buck, A.K., Liu, X., Winkler, M. and Reske, S.N. (2003) Gene silencing by adenovirus-delivered siRNA. *FEBS Lett* **539**: 111–114.

Shinagawa, T. and Ishii, S. (2003) Generation of Ski-knockdown mice by expressing a long double-strand RNA from an RNA polymerase II promoter. *Genes Dev* **17**: 1340–1345.

Silva, J.M., Li, M.Z., Chang, K., *et al.* (2005) Second-generation shRNA libraries covering the mouse and human genomes. *Nat Genet* **37**: 1281–1288.

Sledz, C.A., Holko, M., de Veer, M.J., Silverman, R.H. and Williams, B.R. (2003) Activation of the interferon system by short-interfering RNAs. *Nat Cell Biol* **5**: 834–839.

Sledz, C.A. and Williams, B.R.G. (2004) RNA interference and double-stranded-RNA-activated pathways. *Biochem Soc Trans* **32**: 952–956.

Smart, N., Scambler, P.J. and Riley, P.R. (2005) A rapid and sensitive assay for quantification of siRNA efficiency and specificity. *Biol Proced Online* **7**: 1–7.

Spagnou, S., Miller, A.D. and Keller, M. (2004) Lipidic carriers of siRNA: differences in the formulation, cellular uptake, and delivery with plasmid DNA. *Biochemistry* **43**: 13348–13356.

Stark, G.R., Kerr, I.M., Williams, B.R., Silverman, R.H. and Schreiber, R.D. (1998) How cells respond to interferons. *Annu Rev Biochem* **67**: 227–264.

Wheeler, D.B., Carpenter, A.E. and Sabatini, D.M. (2005) Cell microarrays and RNA interference chip away at gene function. *Nat Genet* **37** Suppl: S25–S30.

Xin, H., Bernal, A., Amato, F.A., *et al.* (2004) High-throughput siRNA-based functional target validation. *J Biomol Screen* **9**: 286–293.

Yao, X., Freas, A., Ramirez, J., Demirev, P.A. and Fenselau, C. (2001) Proteolytic ^{18}O labeling for comparative proteomics: model studies with two serotypes of adenovirus. *Anal Chem* **73**: 2836–2842.

Yuan, Y.R., Pei, Y, Ma, J.B., *et al.* (2005) Crystal structure of *A. aeolicus* argonaute, a site-specific DNA-guided endoribonuclease, provides insights into RISC-mediated mRNA cleavage. *Mol Cell* **19**: 405–419.

Zhang, A., Pastor, L., Nguyen, Q., *et al.* (2005) Small interfering RNA and gene expression analysis using a multiplex branched DNA assay without RNA purification. *J Biomol Screen* **10**: 549–556.

Zhao, H.F., L'Abbe, D., Jolicoeur, N., Wu, M.Q., Li, Z., Yu, Z.B. and Shen, S.H. (2005) High-throughput screening of effective siRNAs from RNAi libraries delivered via bacterial invasion. *Nat Methods* **2**: 967–973.

RNAi libraries in dissecting molecular pathways of the human cell

5

Cheryl Eifert, Antonis Kourtidis and Douglas S. Conklin

5.1 Introduction

Loss-of-function genetic experiments remain one of the most effective ways to gain insight into a gene's function. A significant drawback to this approach is that the construction of knockouts has historically been an arduous process that is not ensured success, since multiple copies of a gene or gene compensation by related genes can mask a phenotype. The advent of RNAi has revolutionized loss-of-function analyses because of its ease of use, the lack of need for any prior information on the biological system being evaluated, its effectiveness in inhibiting all homologous transcripts regardless of how many gene copies are present, and the applicability of its use in large-scale studies. Currently, there is a variety of different RNAi constructs available to generate gene knockdowns, including whole genome RNAi libraries that enable loss-of-function screens on a genome-wide basis. Such screens are unparalleled in their ability to identify factors and pathways that are critical for any given process. As such, these screens are being used to expedite the search for novel, more effective and specific therapeutic targets.

5.2 RNAi

RNAi is a naturally occurring biological process used by nearly all organisms to inhibit either gene expression or protein synthesis. RNAi serves multiple purposes, being initiated from exogenously supplied dsRNAs (viral, pathogenic, etc.) or from endogenously transcribed miRNAs. RNAi was originally demonstrated in the nematode *Caenorhabditis elegans*, where it was shown that the introduction of long dsRNAs that are homologous to an endogenous gene sequence caused the inhibition of that same gene (Fire *et al.*, 1998). It was subsequently shown that the long dsRNAs, were recognized and cleaved into dsRNAs of 21–26 nt in length, designated as siRNAs, by the RNase III containing enzyme complex called Dicer (Bernstein *et al.*, 2001; Zamore *et al.*, 2000). siRNA duplexes are unwound by the RISC, which through base pairing between the siRNA and an endogenous, complementary transcript leads to its degradation (Hammond, 2005; Hammond *et al.*, 2001).

It was later recognized that RNAi was not only a defensive response to foreign nucleic acids but was also a required aspect of normal development. Let-7 and lin-4 are developmentally regulated miRNAs, initially identified in *C. elegans*, whose expression leads to the translational suppression, as well as the degradation of mRNAs that are important for nematode development (Bagga *et al.*, 2005; Grishok *et al.*, 2001; Vella *et al.*, 2004). miRNAs have since been found in vertebrates, including humans, where there are now predicted to be over 300 endogenous miRNAs present in the genome (Griffiths-Jones *et*

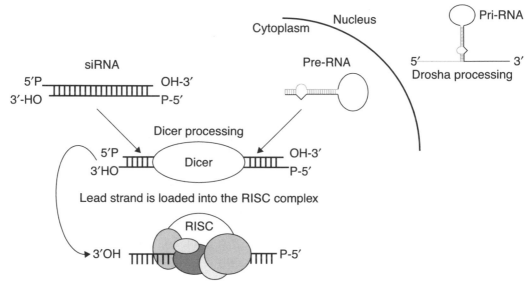

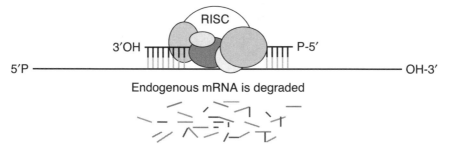

Figure 5.1

RNAi machinery. Libraries comprised of synthetically generated siRNAs or shRNAs mediate RNAi. shRNAs are designed to mimic endogenous miRNAs, which are transcribed in the nucleus and form a conserved stem-loop secondary structure with extended 5′ and 3′ 'tail' sequences (pri-RNAs). The pri-RNA is cleaved by Drosha and exported to the cytoplasm as a shortened pre-RNA (shRNA). The pre-RNA structure is recognized and processed by Dicer into the standard 21–26mer siRNA. Loading of the lead strand into the RISC complex then leads to degradation or translational inhibition of homologous transcripts.

al., 2006; Lewis *et al.*, 2003). Although most information supports the idea that siRNAs and miRNAs utilize the same RNAi machinery to effect gene silencing (Zeng and Cullen, 2003), miRNAs unlike siRNAs, can lead to translational inhibition as well as to transcript degradation (Bagga *et al.*, 2005; Murchison and Hannon, 2004; Valencia-Sanchez *et al.*, 2006).

miRNAs are transcribed as long, primary miRNA transcripts (pri-miRNA), and form conserved stem-loop secondary structures containing 5' and 3' extended, single-stranded, tail sequences. The pri-miRNA structure is recognized and cleaved by an RNase III-containing enzyme complex, named Drosha, into a smaller (65–75 nt) pre-miRNA stem-loop secondary structure. The pre-miRNA is then exported from the nucleus into the cytoplasm where Dicer recognizes and cleaves it into the standard 21–26 nt dsRNA (miRNA) (Han *et al.*, 2006; Lee *et al.*, 2002) (*Figure 5.1*). The information gleaned from the elucidation of endogenous miRNA gene silencing has been used to design RNAi constructs that can be expressed from viral DNA vectors (shRNAs). The shRNA sequences mimic conserved pre-miRNA features and as such are processed by the endogenous RNAi machinery into functional miRNA fragments (Silva *et al.*, 2005).

Both synthetically generated siRNAs and shRNA-containing vectors are currently being used to knock down gene expression and both have been developed into libraries that cover entire genomes, including the human, mouse, rat (etc.) (*Table 5.1*). Libraries containing synthetic siRNAs have the benefit of being experimentally verified to ensure that each targeted gene will be efficiently knocked down. SiRNAs, since they cannot be reproduced, however, are relatively expensive for most academic research groups,

Table 5.1 A summary of mammalian RNAi libraries

Company or group	Coverage	Species	Method or expression vector	Reference
siRNA libraries				
Dharmacon	Genome-wide	Human, mouse, rat	Synthetic oligos	Aza-Blanc *et al.*, 2003
Max Planck	Half the genome	Human	Endonuclease-prepared siRNAs	Kittler *et al.*, 2004
Ambion	Genome-wide	Human, mouse, rat	Synthetic oligos	Ovcharenko *et al.*, 2005
Qiagen	Genome-wide	Human	Synthetic oligos	MacKeigan *et al.*, 2005
Invitrogen	Kinases	Human	Synthetic oligos	–
shRNA libraries				
NKI	One-third of genome	Human	pRetroSuper	Berns *et al.*, 2004
CSHL/Harvard	One-third of genome	Human, mouse	pSHAG-MAGIC 1	Paddison *et al.*, 2004
CSHL/Harvard	Genome-wide	Human, mouse	pSHAG-MAGIC 2	Silva *et al.*, 2005
Univ. of Tokyo	Apoptosis-related genes	Human	piGENE PURhU6	Futami *et al.*, 2005
RNAi consortium	Two-thirds of genome	Human, mouse	pLKO.1-puro	Moffat *et al.*, 2006
NCI	One-tenth of genome	Human	pRSMX	Ngo *et al.*, 2006
miRNA libraries				
NKI	All annotated miRNAs	Human	miR-Vec	Voorhoeve *et al.*, 2006

especially for genome-wide applications. siRNA molecules must be transfected into cells by either electroporation or cationic lipids, neither of which are very efficient *in vivo* and make developmental or long-term studies infeasible in mammalian cells using these methods. To remedy these limitations, a number of groups developed vector-based libraries expressing siRNAs (Chen *et al.*, 2005; Kaykas and Moon, 2004; Zheng *et al.*, 2004) or shRNAs (Berns *et al.*, 2004; Futami *et al.*, 2005; McManus *et al.*, 2002; Paddison *et al.*, 2002, 2004; Shirane *et al.*, 2004; Silva *et al.*, 2005) to enable stable, long-term, RNAi gene silencing.

shRNAs are modeled after pre-miRNAs and are transcribed from a single promoter contained on DNA-based vectors including those with retroviral (Brummelkamp *et al.*, 2002; Paddison *et al.*, 2004; Shirane *et al.*, 2004; Silva *et al.*, 2005), lentiviral (Bailey *et al.*, 2006; Morris and Rossi, 2006; Rubinson *et al.*, 2003; Stegmeier *et al.*, 2005; Ventura *et al.*, 2004) or adenoviral (Carette *et al.*, 2004; Chen *et al.*, 2006; Li *et al.*, 2005; Shen and Reske, 2004) backbones. Vectors expressing hairpins allow for the continuous production of siRNAs. Since these vectors can be stably integrated into the genome, the enrichment of cells containing the silencing construct, continuity through passage, storage as frozen stocks, and production of silencing-based transgenics that include passage through the germline are all possible (Carmell *et al.*, 2003; Szulc *et al.*, 2006).

The recent demonstration that pol II promoters drive the expression of endogenous miRNAs (Cai *et al.*, 2004; Lee *et al.*, 2004) has been incorporated in the design of pol II driven shRNA vectors (Dickins *et al.*, 2005; Stegmeier *et al.*, 2005; Zhou *et al.*, 2005). Notably the use of pol II promoters has permitted the development of regulatable (e.g. tetracycline inducible) shRNA vectors (Chang *et al.*, 2004; Czauderna *et al.*, 2003; Dickins *et al.*, 2005; Matthess *et al.*, 2005; Stegmeier *et al.*, 2005; Szulc *et al.*, 2006; Yang *et al.*, 2005). This has enabled controlled gene knockdown experiments, as well as the analysis of lethal gene knockdowns.

Finally, shRNA vector libraries incorporate 'barcodes', which are unique ~60-mer sequences attached to individual shRNA vectors, enabling the identification of depleted or over-represented shRNAs using DNA microarrays (Berns *et al.*, 2004; Paddison *et al.*, 2004; Westbrook *et al.*, 2005). Genomic DNA isolated from cells infected with a pool of shRNA constructs containing barcodes is amplified using a primer pair that is just outside the barcode region and complementary to the vector. The PCR products from control cells and experimental cells are then labeled with two different fluorescent dyes, which are subsequently hybridized to a DNA microarray that contains complementary oligonucleotides. The microarray is then scanned to reveal shRNAs that have either been depleted or gained in the experimental population relative to the control population. This type of screen is particularly useful for determining genes whose inhibition alone or in combination with another shRNA (synthetic lethality) causes cell death (i.e. depletion of the shRNA from the pool).

5.3 Approaches for loss-of-function screens

A combination of different approaches has been employed in the use of RNAi to perform functional screens in human cells (*Figure 5.2*). RNAi screens can be performed either using arrays of individual RNAi constructs in high-throughput screening protocols, or by using pools of shRNA constructs. Arrayed RNAi screens utilize any number of high-throughput assays for phenotypic selection while the pooled approach depends upon a selective phenotype. The pooled-shRNA approach is preferable to siRNA arrays when cost is an issue; shRNA vectors can be easily reproduced in bacterial hosts and used repeatedly, contrary to commercially available synthetic oligos, and pooled RNAi screens rely upon phenotypic selection, which typically avoids the use of costly, high-throughput mode assays that can become inhibitory for genome-wide analyses. On the other hand, vector-expressed shRNAs are less efficient than siRNAs at conferring gene knockdown, whereas the pooled-RNAi approach also requires a second round of phenotypic selection and increases the possibility of off-target effects. Nevertheless, a number of studies have proved the effectiveness and applicability of this approach while revealing important components of

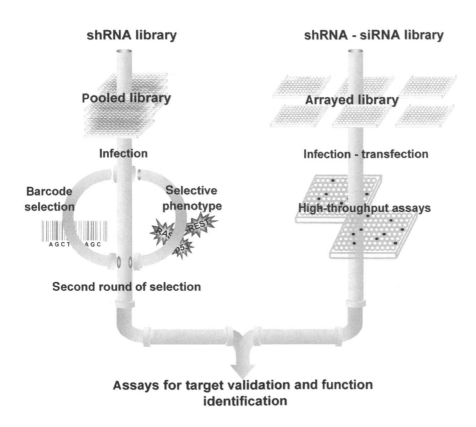

Figure 5.2

Major RNAi library approaches add high-throughput target identification and validation to the pipeline.

cellular mechanisms (Berns *et al.*, 2004; Brummelkamp *et al.*, 2006; Ngo *et al.*, 2006; Westbrook *et al.*, 2005).

Additionally, RNAi screens may be designed to target only specific functional subsets of the genome or to interrogate the entire or a significant portion of the genome. Choosing the appropriate RNAi library for any given screen is paramount, since the appropriate library can limit the time and cost of the study without compromising the robustness of the results. Selection of the appropriate RNAi library is determined by the question at hand. For instance, certain projects seek to elucidate the role of a specific group of genes, such as kinases, in a particular human disease, such as cancer. In this case, a hairpin library targeting only the kinases would be desirable. Furthermore, selection of a specific RNAi library subset will contribute to the screen by dictating the type of phenotypic selection (based upon a specific function or mechanism under study), as well as aiding in the interpretation of biologically relevant results. When novel factors are sought in a particular process or pathway, however, a genomic library may be required. A description of large-scale RNAi screens performed to date is presented in *Table 5.2*.

5.4 High-throughput RNAi screens

High-throughput screens using siRNAs or shRNAs arrayed in multi-well format target one gene at a time in the host cell. The first such screen performed in human cells sought to identify modulators of the TRAIL-induced apoptotic pathway (Aza-Blanc *et al.*, 2003). TRAIL is a member of the TNF family, which has been shown to selectively activate apoptosis in tumor cells. The screen utilized siRNAs targeting 510 human genes, including 380 known and predicted kinases and a hundred genes with unknown function. HeLa cells were transfected with the library, either with or without TRAIL, and relative proliferation levels were determined using a redox indicator dye (alamarBlue; Biosource). Using this approach, both sensitizers and inhibitors of TRAIL-mediated apoptosis were identified, including genes not previously associated with TRAIL-mediated apoptosis, such as GSK3α, SRP72, FLJ32312, PAK1, JIK and MIRSA.

To ensure that the effect of the siRNAs was specific rather than the result of off-target effects, cells were transfected with an additional siRNA construct targeting each newly identified gene. A caspase-3/7 detection assay confirmed that GSK3α, SRP72, and FLJ32312 have pro-apoptotic roles while PAK1, JIK, and MIRSA each blocked TRAIL-mediated apoptosis.

Interestingly, a subsequent RNAi screen also addressed the role of kinases in HeLa cells (Pelkmans *et al.*, 2005) but focused on two principal types of endocytosis – clathrin- and caveolae/raft-mediated. To elucidate the role of kinases in endocytosis, the authors transfected cells with a siRNA library targeting 590 kinases and monitored the subsequent infection efficiency of the vesicular stomatitis virus (VSV) and the simian virus 40 (SV40), which use the clathrin and caveolae/raft endocytic routes to invade human cells, respectively. This screen revealed the previously unrecognized magnitude of kinase involvement in endocytosis. Two hundred and ten kinases were found to impact the two distinct modes of endocytosis with 47 of them being either poorly or entirely uncharacterized genes. To organize the genes

Table 5.2 RNAi functional genomic screens

Title	Reference
C. elegans	
Chromosome I genes	Fraser *et al.*, 2000
Chromosome III genes	Gonczy *et al.*, 2000
Ovary	Piano *et al.*, 2000
Embryogenesis	Maeda *et al.*, 2001
Protection against mutations	Pothof *et al.*, 2003
Metabolism	Ashrafi *et al.*, 2003
Transposon silencing	Vastenhouw *et al.*, 2003
Polyglutamine aggregation	Nollen *et al.*, 2004
RNA interference mechanism	Kim *et al.*, 2005
Early embryogenesis	Sonnichsen *et al.*, 2005
Sumoylation	Poulin *et al.*, 2005
Longevity	Hamilton *et al.*, 2005
Synapse structure and function	Sieburth *et al.*, 2005
Molting cycle	Frand *et al.*, 2005
Lifespan	Hansen *et al.*, 2005
Drosophila	
Morphogenesis	Kiger *et al.*, 2003
Hedgehog signaling	Lum *et al.*, 2003
Cell growth and viability	Boutros *et al.*, 2004
Innate immune response	Foley and O'Farrell, 2004
Nervous system development	Ivanov *et al.*, 2004
CD36 family member required for mycobacterial infection	Nybakken *et al.*, 2005
RNA virus sensitivity	Cherry *et al.*, 2005
Kinase substrates	Lee *et al.*, 2005
Wnt-wingless signaling	DasGupta *et al.*, 2005
Store-operated Ca^{2+} channel function	Roos *et al.*, 2005
Intracellular bacterial infection	Agaisse *et al.*, 2005
Mycobacterial infection	Philips *et al.*, 2005
JAK/STAT signaling	Baeg *et al.*, 2005
JAK/STAT signaling	Muller *et al.*, 2005
Imd signaling	Kleino *et al.*, 2005
Apoptosis	Gesellchen *et al.*, 2005
Toll pathway	Kambris *et al.*, 2006
Protein secretion – Golgi organization	Bard *et al.*, 2006
Neuron dendrite development	Parrish *et al.*, 2006
TOR-regulated genes	Guertin *et al.*, 2006
NFAT regulators	Gwack *et al.*, 2006
Identification of genes that regulate Ca^{2+} channels	Zhang *et al.*, 2006
Mammals	
TRAIL-induced apoptosis	Aza-Blanc *et al.*, 2003
Endocytosis	Pelkmans *et al.*, 2005
p53-dependent proliferation arrest	Berns *et al.*, 2004
Proteosome function	Silva *et al.*, 2004b
Cell division	Kittler *et al.*, 2004
PI3K pathway	Hsieh *et al.*, 2004
ER stress-induced apoptosis	Matsumoto *et al.*, 2005
Apoptosis and chemoresistance	MacKeigan *et al.*, 2005
Tumor suppressors	Westbrook *et al.*, 2005
Ras tumorigenicity	Nicke *et al.*, 2005
Breast cancer tumor suppression	Kolfschoten *et al.*, 2005
Akt-related kinases	Morgan-Lappe *et al.*, 2006
B-cell lymphoma oncogenes	Ngo *et al.*, 2006
miRNAs related to oncogenesis	Voorhoeve *et al.*, 2006

into functionally relevant groups, a two-step cluster analysis was conducted. These two works demonstrate the power of RNAi screening; though both groups targeted the same set of genes in the same cell line, they were able to address unique biological questions by employing different phenotypic selections.

shRNA libraries have also been used for high-throughput screens. An arrayed screen performed using an shRNA library designed in a retroviral vector (MSCV) and transcribed from a U6 pol III type promoter (pSHAG-MAGIC) identified several proteasome components, indicating that a substantial percentage of the constructs were functional in a biologically relevant context (Paddiston *et al.*, 2004). Another screen identified several new modulators of the p53 pathway (Barns *et al.*, 2004). Although these successful studies were carried out using transient transfections of simple shRNA constructs, second-generation shRNA libraries incorporating features of endogenous miRNA silencing and novel vector designs are now available (Stegmeier *et al.*, 2005).

5.5 RNAi-induced phenotype selections

Selective phenotypic screens generally require the wholesale introduction of RNAi library constructs into cells. Those cells that exhibit the selected phenotype are then harvested so that the responsible shRNA vector can be identified by sequencing or using a DNA bar-code microarray. Berns *et al.* (2004) describe a selective phenotypic RNAi screen in human cells that led to the identification of new components of the p53 pathway. The screen utilizes a retroviral library that encodes 23 742 shRNAs targeting 7914 distinct human genes. To identify new components of the p53 pathway, a cell system was devised to screen for bypass of p53-dependent proliferation arrest. Primary human BJ fibroblast cells were engineered to express the murine ecotropic receptor, the telomerase catalytic subunit (TERT) and a temperature-sensitive allele of SV40 large T-antigen (tsLT) (BJ-TERT-tsLT cells). These cells proliferate at 32°C but arrest at 39°C.

To perform the large-scale loss-of-function screen, the shRNA library was pooled and retroviral supernatants were used to infect BJ-TERT-tsLT cells. Colonies that were able to proliferate at the non-permissive temperature were recovered and the shRNA plasmids within were isolated for sequencing. All together six genes were identified whose knockdown could evade the temperature shift-induced growth arrest in BJ-TERT-tsLT cells, including p53 itself. The five additionally identified genes were the ribosomal S6 kinase 4 (RSK4), histone acetyl transferase TIP60 (HTATIP), histone deacetylase 4 (HDAC4), a putative S-adenosyl-L-homocysteine hydrolase, SAH3 (KIAA0828), and T-complex protein 1, β-subunit (CCT2). To verify the ability to bypass p53-dependent growth arrest was due specifically to each of the five newly identified genes rather than to an off-target effect an additional shRNA construct, targeting each gene was shown to confer the knockdown of each cognate and endogenous gene and, moreover, was able to confer escape from temperature shift-induced growth arrest in BJ-TERT-tsLT cells.

The first example of a selective phenotypic screen that was used in conjunction with a DNA bar-code microarray screen to identify shRNA constructs within selected cells focused on the inactivation of genes that inhibited anchorage-independent growth (Westbrook *et al.*, 2005). The pSHAG-MAGIC1 shRNA retroviral library, consisting of ~28 000 shRNAs targeting ~9000 genes, was pooled and used to infect human mammary epithelial cells that were immortalized with hTERT and the large T-antigen (TLM-HMEC). To isolate tumor suppressor genes, shRNA-infected cells were incubated in semisolid media and monitored for anchorage-independent growth. Approximately 100 shRNA-infected colonies grew in an anchorage-independent manner indicating transformation. A barcode microarray was then used to monitor enriched shRNAs in the pooled population of genomic DNAs. A number of well-established tumor suppressors were identified, such as PTEN and TGFBR2, as well as a new candidate tumor suppressor, the RE1-silencing transcription factor, REST/NRSF. These findings were, subsequently, confirmed in experiments, which over- and under-expressed REST. Additionally, a genomic analysis of cancer cells was performed revealing that REST is, in fact, absent from colon cancers. This study effectively demonstrated the potential and cost effectiveness of using a pooled shRNA screen in conjunction with barcode monitoring for the identification of novel tumor suppressors.

Ngo *et al.* (2006) used the selective phenotypic approach to define genes that are required for the proliferation and survival of one subtype of cancer cells but not for another closely related subtype. To better enable the identification of genes whose knockdown is lethal, they created an shRNA library in a doxycycline-inducible retroviral vector. The loss-of-function screen consisted of 1854 shRNA vectors targeting 683 human genes. Two molecularly distinct subtypes of diffuse large B-cell lymphoma (DLBCL) cells, activated B-cell-like DLBCLs and germinal center B-cell-like DLBCLs (Alizadeh *et al.*, 2000) were transduced in pools from the retroviral library. A DNA barcode microarray was then used to identify shRNAs that had been selectively removed from doxycycline-induced populations. Altogether, 17 shRNA vectors targeting 15 genes were significantly depleted from at least two cell lines. Of particular interest were shRNAs targeting four genes that were specifically toxic to activated B-cell-like DLBCLs but not germinal center B-cell-like DLBCLs. All four selectively toxic genes targeted components of the NFκB pathway, including IKBKB, CARD11, MALT1, and BCL10.

To provide additional evidence that only activated B-cell-like DLBCLs require CARD11 for survival, a shRNA targeting CARD11 was cloned into a vector coexpressing GFP, and FACS analysis was used to monitor GFP expression. In activated B-cell-like DLBCL cell lines, the level of GFP dropped over time indicating cellular toxicity, whereas no change in cellular viability was seen in germinal center B-cell-like DLBCL cell lines. To address whether the identified genes were participating in the NFκB pathway, the *Photinus* luciferase protein was fused to the IkBa protein and luciferase activity was assessed in both DLBCL subtypes containing shRNAs targeting either CARD11 or MALT1. Luciferase activity increased in activated B-cell like DLBCL cells transduced with either CARD11 or MALT1 but did not in germinal center B-cell-like DLBCL cells, indicating that only activated B-cell like DLBCL cells require NFκB signaling for survival. To further establish that CARD11 is involved in the NFκB pathway, downstream targets of the NFκB pathway were evaluated using a microarray analysis. NFκB target genes were significantly downregulated in the activated B-cell-like DLBCL cell line that was transduced with an shRNA targeting CARD11, verifying that CARD11 participates in the NFκB pathway. This study demonstrated the capacity of such loss-of-function RNAi screens to generate a functional taxonomy of cancer subtypes that will promote the identification of new and highly specific therapeutic targets.

5.6 Screens for miRNA functions

miRNAs comprise a large family of endogenously expressed, non-coding genes. More than 300 members have already been identified and a genome-wide sequence analysis, using bioinformatic algorithms, indicates that they may represent ~4% of the human genome (Griffiths-Jones *et al.*, 2006; Jones-Rhoades and Bartel, 2004). A number of miRNAs have been experimentally shown to regulate a variety of target mRNAs (Lim *et al.*, 2005; O'Donnell *et al.*, 2005; Schratt *et al.*, 2006) and several differentially expressed miRNAs have been associated with cancer (He *et al.*, 2005; Lu *et al.*, 2005). Unfortunately, the function of only a handful of miRNAs has been experimentally determined. In an attempt to assign functional roles to miRNAs, Voorhoeve *et al.* (2006) performed a loss-of-function screen using an pMSCV retroviral RNAi library (miR-Vec) that was constructed from a majority of annotated human miRNAs (miR-Lib). In addition, a complementary microarray (miR-Array) consisting of all the PCR-amplified library inserts was created.

Previous studies have shown that primary BJ fibroblasts containing an SV-small t antigen and hTERT (BJ-ET) will undergo p53-dependent premature senescence in the presence of an oncogene, such as RAS^{V12}. Tumorigenicity is impeded by this mechanism as loss of p53-dependent processes are necessary for transformation (Voorhoeve and Agami, 2003). To identify miRNAs whose depletion can overcome p53-dependent premature senescence, BJ-ET cells were infected with the miR-Lib followed by the RAS^{V12} oncogene. The abundance of the miR-Vecs was, subsequently, monitored in the cell population using the miR-Array. The results of the array analysis indicated that depletion of two of the miRNAs, miR-372 or miR-373, was sufficient to enable escape from p53-mediated growth arrest. To determine if the miR-371–3 cluster had any significance in human cancers, a number of testicular germ cell tumor (TGCT) cell lines were evaluated for the status of the miR-371–3 cluster, since these tumors typically retain wild-type p53 protein. Importantly, they found that the miR-371–3 cluster was expressed in four of seven TGCT cell lines, whereas it was absent from all somatic cell lines tested, suggesting that these miRNAs function downstream of p53 in the pathway. To strengthen the connection to the p53 pathway cells that do not contain a functional p53 and do not express miR-371–3 were transfected with, first, p21-RFP, which resulted in cell-cycle arrest, and next with the miR-371–3 cluster, which resulted in continued cellular proliferation.

To identify potential targets of the miR-371–3 cluster, a microarray analysis was used to compare the expression profile of genes from RAS^{V12} BJ/ET cells containing either the miR-372/3 or p53^{kd}. The result of the array analysis used in conjunction with a miRNA target prediction program pointed to three potential targets: the Large Tumor Suppressor homologue 2 (LATS2), FYVE and coiled coil containing protein 1 (FYCO1), and (Suv39-H1). LATS2 was chosen to proceed with because, among other indicators, previous results indicated that LATS2 was indeed involved in the inhibition of cyclin E/CDK2 as well as human cancers. In fact, they found that the miR-371–3 cluster caused a twofold decrease in LATS2 gene expression and a four- to fivefold decrease in protein levels.

This was a groundbreaking work because it was the first study to success-fully assign a specific function to a known human miRNA. The approach is similar to gain-of-function, cDNA screens performed in the past, except miRNAs may be more amenable to targeted therapeutics. Chemically engineered oligonucleotides, called 'antagomirs' have recently been used to silence miRNAs *in vivo* (Krutzfeldt *et al.*, 2005), providing an additional tool for loss-of-function studies on miRNAs. The loss-of-function arsenal currently includes siRNAs, shRNAs, miRNAs, and antagomirs, which will no doubt dramatically accelerate the assignation of function to both genes and miRNAs and in the process elucidate the networks that coordinate the multitude of cellular behaviors.

5.7 Perspectives in disease treatment

The major challenge that has emerged following the completion of the genome sequencing projects was not only to discern gene function but also to formulate a comprehensive view of cellular regulation such that key nodes for therapeutic intervention can be identified. Functional genomics using RNAi has facilitated the rate at which functions are assigned to genes and has thereby expedited the identification of potential therapeutic targets for a number of diseases (Hannon and Rossi, 2004; Ito *et al.*, 2005; Silva *et al.*, 2004a). A more immediate impact of RNAi in disease treatment may result from its deployment as a therapeutic (Stevenson, 2004). RNAi constructs can be designed to target any known gene and are a particularly effective means to inhibit aberrant gene expression that result in pathogen-esis. The mechanism of RNAi provides the high specificity required for targeted therapies, potentially overcoming the side-effects of several thera-pies already in use. Although obstacles to its direct use in patients remain, such as delivery methods and off-target silencing (Mocellin *et al.*, 2006; Ryther *et al.*, 2005; Stevenson, 2004), several clinical trials are now under-way. The hope is that RNAi will not simply remain a powerful research tool, but that it will also become an ideal therapeutic for virtually any disease.

References

Agaisse, H., Burrack, L.S., Philips, J.A., Rubin, E.J., Perrimon, N. and Higgins, D.E. (2005) Genome-wide RNAi screen for host factors required for intracellular bacterial infection. *Science* **309**: 1248–1251.

Alizadeh, A.A., Eisen, M.B., Davis, R.E., *et al.* (2000) Distinct types of diffuse large B-cell lymphoma identified by gene expression profiling. *Nature* **403**(6769): 503–511.

Ashrafi, K., Chang, F.Y., Watts, J.L., Fraser, A.G., Kamath, R.S., Ahringer, J. and Ruvkun, G. (2003) Genome-wide RNAi analysis of *Caenorhabditis elegans* fat regulatory genes. *Nature* **421**: 268–272.

Aza-Blanc, P., Cooper, C.L., Wagner, K., Batalov, S., Deveraux, Q.L. and Cooke, M.P. (2003) Identification of modulators of TRAIL-induced apoptosis via RNAi-based phenotypic screening. *Mol Cell* **12**: 627–637.

Baeg, G.H., Zhou, R. and Perrimon, N. (2005) Genome-wide RNAi analysis of JAK/STAT signaling components in *Drosophila*. *Genes Dev* **19**: 1861–1870.

Bagga, S., Bracht, J., Hunter, S., Massirer, K., Holtz, J., Eachus, R. and Pasquinelli, A.E. (2005) Regulation by let-7 and lin-4 miRNAs results in target mRNA degradation. *Cell* **122**: 553–563.

Bailey, S.N., Ali, S.M., Carpenter, A.E., Higgins, C.O. and Sabatini, D.M. (2006) Microarrays of lentiviruses for gene function screens in immortalized and primary cells. *Nat Methods* **3**: 117–122.

Bard, F., Casano, L., Mallabiabarrena, A., *et al.* (2006) Functional genomics reveals genes involved in protein secretion and Golgi organization. *Nature* **439**: 604–607.

Berns, K., Hijmans, E.M., Mullenders, J., *et al.* (2004) A large-scale RNAi screen in human cells identifies new components of the p53 pathway. *Nature* **428**: 431–437.

Bernstein, E., Caudy, A.A., Hammond, S.M. and Hannon, G.J. (2001) Role for a bidentate ribonuclease in the initiation step of RNA interference. *Nature* **409**: 363–366.

Boutros, M., Kiger, A.A., Armknecht, S., Kerr, K., Hild, M., Koch, B., Haas, S.A., Paro, R. and Perrimon, N. (2004) Genome-wide RNAi analysis of growth and viability in *Drosophila* cells. *Science* **303**: 832–835.

Brummelkamp, T.R., Bernards, R. and Agami, R. (2002) Stable suppression of tumorigenicity by virus-mediated RNA interference. *Cancer Cell* **2**: 243–7.

Brummelkamp, T.R., Fabius, A.W., Mullenders, J., Madiredjo, M., Velds, A., Kerkhoven, R.M., Bernards, R. and Beijersbergen, R.L. (2006) An shRNA barcode screen provides insight into cancer cell vulnerability to MDM2 inhibitors. *Nat Chem Biol* **2**: 202–206.

Cai, X., Hagedorn, C.H. and Cullen, B.R. (2004) Human microRNAs are processed from capped, polyadenylated transcripts that can also function as mRNAs. *RNA* **10**: 1957–1966.

Carette, J.E., Overmeer, R.M., Schagen, F.H., Alemany, R., Barski, O.A., Gerritsen, W.R. and Van Beusechem, V.W. (2004) Conditionally replicating adenoviruses expressing short hairpin RNAs silence the expression of a target gene in cancer cells. *Cancer Res* **64**: 2663–2667.

Carmell, M.A., Zhang, L., Conklin, D.S., Hannon, G.J. and Rosenquist, T.A. (2003) Germline transmission of RNAi in mice. *Nat Struct Biol* **10**: 91–92.

Chang, H.S., Lin, C.H., Chen, Y.C. and Yu, W.C. (2004) Using siRNA technique to generate transgenic animals with spatiotemporal and conditional gene knockdown. *Am J Pathol* **165**: 1535–1541.

Chen, M., Zhang, L., Zhang, H.Y., Xiong, X., Wang, B., Du, Q., Lu, B., Wahlestedt, C. and Liang, Z. (2005) A universal plasmid library encoding all permutations of small interfering RNA. *Proc Natl Acad Sci USA* **102**: 2356–2361.

Chen, Y., Chen, H., Hoffmann, A., Cool, D.R., Diz, D.I., Chappell, M.C., Chen, A. and Morris, M. (2006) Adenovirus-mediated small-interference RNA for *in vivo* silencing of angiotensin AT1a receptors in mouse brain. *Hypertension* **47**: 230–237.

Cherry, S., Doukas, T., Armknecht, S., Whelan, S., Wang, H., Sarnow, P. and Perrimon, N. (2005) Genome-wide RNAi screen reveals a specific sensitivity of IRES-containing RNA viruses to host translation inhibition. *Genes Dev* **19**: 445–452.

Czauderna, F., Santel, A., Hinz, M., *et al.* (2003) Inducible shRNA expression for application in a prostate cancer mouse model. *Nucleic Acids Res* **31**: e127.

DasGupta, R., Kaykas, A., Moon, R.T. and Perrimon, N. (2005) Functional genomic analysis of the Wnt-wingless signaling pathway. *Science* **308**: 826–833.

Dickins, R.A., Hemann, M.T., Zilfou, J.T., Simpson, D.R., Ibarra, I., Hannon, G.J. and Lowe, S.W. (2005) Probing tumor phenotypes using stable and regulated synthetic microRNA precursors. *Nat Genet* **37**: 1289–1295.

Fire, A., Xu, S., Montgomery, M.K., Kostas, S.A., Driver, S.E. and Mello, C.C. (1998) Potent and specific genetic interference by double-stranded RNA in *Caenorhabditis elegans*. *Nature* **391**: 806–811.

Foley, E. and O'Farrell, P.H. (2004) Functional dissection of an innate immune response by a genome-wide RNAi screen. *PLoS Biol* **2**: E203.

Frand, A.R., Russel, S. and Ruvkun, G. (2005) Functional genomic analysis of *C. elegans* molting. *PLoS Biol* **3**: e312.

Fraser, A.G., Kamath, R.S., Zipperlen, P., Martinez-Campos, M., Sohrmann, M. and Ahringer, J. (2000) Functional genomic analysis of *C. elegans* chromosome I by systematic RNA interference. *Nature* **408**: 325–330.

Futami, T., Miyagishi, M. and Taira, K. (2005) Identification of a network involved in thapsigargin-induced apoptosis using a library of small interfering RNA expression vectors. *J Biol Chem* **280**: 826–831.

Gesellchen, V., Kuttenkeuler, D., Steckel, M., Pelte, N. and Boutros, M. (2005) An RNA interference screen identifies inhibitor of apoptosis protein 2 as a regulator of innate immune signalling in *Drosophila*. *EMBO Rep* **6**: 979–984.

Gonczy, P., Echeverri, C., Oegema, K., *et al.* (2000) Functional genomic analysis of cell division in *C. elegans* using RNAi of genes on chromosome III. *Nature* **408**: 331–336.

Griffiths-Jones, S., Grocock, R.J., van Dongen, S., Bateman, A. and Enright, A.J. (2006) miRBase: microRNA sequences, targets and gene nomenclature. *Nucleic Acids Res* **34**(Database issue): D140-4.

Grishok, A., Pasquinelli, A.E., Conte, D., *et al.* (2001) Genes and mechanisms related to RNA interference regulate expression of the small temporal RNAs that control *C. elegans* developmental timing. *Cell* **106**: 23–34.

Guertin, D.A., Guntur, K.V., Bell, G.W., Thoreen, C.C. and Sabatini, D.M. (2006) Functional genomics identifies TOR-regulated genes that control growth and division. *Curr Biol* **16**: 958–970.

Gwack, Y., Sharma, S., Nardone, J., *et al.* (2006) A genome-wide *Drosophila* RNAi screen identifies DYRK-family kinases as regulators of NFAT. *Nature* **441**: 646–650.

Hamilton, B., Dong, Y., Shindo, M., Liu, W., Odell, I., Ruvkun, G. and Lee, S.S. (2005) A systematic RNAi screen for longevity genes in *C. elegans*. *Genes Dev* **19**: 1544–1555.

Hammond, S.M. (2005) Dicing and slicing: the core machinery of the RNA interference pathway. *FEBS Lett* **579**: 5822–5829.

Hammond, S.M., Boettcher, S., Caudy, A.A., Kobayashi, R. and Hannon, G.J. (2001) Argonaute2, a link between genetic and biochemical analyses of RNAi. *Science* **293**: 1146–1150.

Han, J., Lee, Y., Yeom, K.H., *et al.* (2006) Molecular basis for the recognition of primary microRNAs by the Drosha-DGCR8 complex. *Cell* **125**: 887–901.

Hannon, G.J. and Rossi, J.J. (2004) Unlocking the potential of the human genome with RNA interference. *Nature* **431**: 371–378.

Hansen, M., Hsu, A.L., Dillin, A. and Kenyon, C. (2005) New genes tied to endocrine, metabolic, and dietary regulation of lifespan from a *Caenorhabditis elegans* genomic RNAi screen. *PLoS Genet* **1**: 119–128.

He, L., Thomson, J.M., Hemann, M.T., *et al.* (2005) A microRNA polycistron as a potential human oncogene. *Nature* **435**: 828–833.

Hsieh, A.C., Bo, R., Manola, J., Vazquez, F., Bare, O., Khvorova, A., Scaringe, S. and

Sellers, W.R. (2004) A library of siRNA duplexes targeting the phosphoinositide 3-kinase pathway: determinants of gene silencing for use in cell-based screens. *Nucleic Acids Res* **32**: 893–901.

Ito, M., Kawano, K., Miyagishi, M. and Taira, K. (2005) Genome-wide application of RNAi to the discovery of potential drug targets. *FEBS Lett* **579**: 5988–5995.

Ivanov, A.I., Rovescalli, A.C., Pozzi, P., *et al.* (2004) Genes required for *Drosophila* nervous system development identified by RNA interference. *Proc Natl Acad Sci USA* **101**: 16216–16221.

Jones-Rhoades, M.W. and Bartel, D.P. (2004) Computational identification of plant microRNAs and their targets, including a stress-induced miRNA. *Mol Cell* **14**: 787–799.

Kambris, Z., Brun, S., Jang, I.H., Nam, H.J., Romeo, Y., Takahashi, K., Lee, W.J., Ueda, R. and Lemaitre, B. (2006) *Drosophila* immunity: a large-scale *in vivo* RNAi screen identifies five serine proteases required for Toll activation. *Curr Biol* **16**: 808–813.

Kaykas, A. and Moon, R.T. (2004) A plasmid-based system for expressing small interfering RNA libraries in mammalian cells. *BMC Cell Biol* **5**: 16.

Kiger, A.A., Baum, B., Jones, S., Jones, M.R., Coulson, A., Echeverri, C. and Perrimon, N. (2003) A functional genomic analysis of cell morphology using RNA interference. *J Biol* **2**: 27.

Kim, J.K., Gabel, H.W., Kamath, R.S., *et al.* (2005) Functional genomic analysis of RNA interference in *C. elegans*. *Science* **308**: 1164–1167.

Kittler, R., Putz, G., Pelletier, L., *et al.* (2004) An endoribonuclease-prepared siRNA screen in human cells identifies genes essential for cell division. *Nature* **432**: 1036–1040.

Kleino, A., Valanne, S., Ulvila, J., *et al.* (2005) Inhibitor of apoptosis 2 and TAK1-binding protein are components of the *Drosophila* Imd pathway. *EMBO J* **24**: 3423–3434.

Kolfschoten, I.G., van Leeuwen, B., Berns, K., Mullenders, J., Beijersbergen, R.L., Bernards, R., Voorhoeve, P.M. and Agami, R. (2005) A genetic screen identifies PITX1 as a suppressor of RAS activity and tumorigenicity. *Cell* **121**: 849–858.

Krutzfeldt, J., Rajewsky, N., Braich, R., Rajeev, K.G., Tuschl, T., Manoharan, M. and Stoffel, M. (2005) Silencing of microRNAs *in vivo* with 'antagomirs'. *Nature* **438**: 685–689.

Lee, L.A., Lee, E., Anderson, M.A., *et al.* (2005) *Drosophila* genome-scale screen for PAN GU kinase substrates identifies Mat89Bb as a cell cycle regulator. *Dev Cell* **8**: 435–442.

Lee, Y., Jeon, K., Lee, J.T., Kim, S. and Kim, V.N. (2002) MicroRNA maturation: stepwise processing and subcellular localization. *EMBO J* **21**: 4663–4670.

Lee, Y., Kim, M., Han, J., Yeom, K.H., Lee, S., Baek, S.H. and Kim, V.N. (2004) MicroRNA genes are transcribed by RNA polymerase II. *EMBO J* **23**: 4051–4060.

Lewis, B.P., Shih, I.H., Jones-Rhoades, M.W., Bartel, D.P. and Burge, C.B. (2003) Prediction of mammalian microRNA targets. *Cell* **115**: 787–798.

Li, H., Fu, X., Chen, Y., Hong, Y., Tan, Y., Cao, H., Wu, M. and Wang, H. (2005) Use of adenovirus-delivered siRNA to target oncoprotein p28GANK in hepatocellular carcinoma. *Gastroenterology* **128**: 2029–2041.

Lim, L.P., Lau, N.C., Garrett-Engele, P., Grimson, A., Schelter, J.M., Castle, J., Bartel, D.P., Linsley, P.S. and Johnson, J.M. (2005) Microarray analysis shows that some microRNAs downregulate large numbers of target mRNAs. *Nature* **433**: 769–773.

Lu, J., Getz, G., Miska, E.A., *et al.* (2005) MicroRNA expression profiles classify human cancers. *Nature* **435**: 834–838.

Lum, L., Yao, S., Mozer, B., Rovescalli, A., Von Kessler, D., Nirenberg, M. and Beachy, P.A. (2003) Identification of Hedgehog pathway components by RNAi in *Drosophila* cultured cells. *Science* **299**: 2039–2045.

MacKeigan, J.P., Murphy, L.O. and Blenis, J. (2005) Sensitized RNAi screen of human kinases and phosphatases identifies new regulators of apoptosis and chemore-sistance. *Nat Cell Biol* **7**: 591–600.

Maeda, I., Kohara, Y., Yamamoto, M. and Sugimoto, A. (2001) Large-scale analysis of gene function in *Caenorhabditis elegans* by high-throughput RNAi. *Curr Biol* **11**: 171–176.

Matsumoto, S., Miyagishi, M., Akashi, H., Nagai, R. and Taira, K. (2005) Analysis of double-stranded RNA-induced apoptosis pathways using interferon-response noninducible small interfering RNA expression vector library. *J Biol Chem* **280**: 25687–25696.

Matthess, Y., Kappel, S., Spankuch, B., Zimmer, B., Kaufmann, M. and Strebhardt, K. (2005) Conditional inhibition of cancer cell proliferation by tetracycline-responsive, H1 promoter-driven silencing of PLK1. *Oncogene* **24**: 2973–2980.

McManus, M.T., Petersen, C.P., Haines, B.B., Chen, J. and Sharp, P.A. (2002) Gene silencing using micro-RNA designed hairpins. *RNA* **8**: 842–850.

Mocellin, S., Costa, R. and Nitti, D. (2006) RNA interference: ready to silence cancer? *J Mol Med* **84**: 4–15.

Moffat, J., Grueneberg, D.A., Yang, X., *et al.* (2006) A lentiviral RNAi library for human and mouse genes applied to an arrayed viral high-content screen. *Cell* **124**(6): 1283–1298.

Morgan-Lappe, S., Woods, K.W., Li, Q., Anderson, M.G., Schurdak, M.E., Luo, Y., Giranda, V.L., Fesik, S.W. and Leverson, J.D. (2006) RNAi-based screening of the human kinome identifies Akt-cooperating kinases: a new approach to designing efficacious multitargeted kinase inhibitors. *Oncogene* **25**: 1340–1348.

Morris, K.V. and Rossi, J.J. (2006) Lentiviral-mediated delivery of siRNAs for antiviral therapy. *Gene Ther* **17**: 479–486.

Muller, P., Kuttenkeuler, D., Gesellchen, V., Zeidler, M.P. and Boutros, M. (2005) Identification of JAK/STAT signalling components by genome-wide RNA inter-ference. *Nature* **436**: 871–875.

Murchison, E.P. and Hannon, G.J. (2004) miRNAs on the move: miRNA biogenesis and the RNAi machinery. *Curr Opin Cell Biol* **16**: 223–229.

Ngo, V.N., Davis, R.E., Lamy, L., *et al.* (2006) A loss-of-function RNA interference screen for molecular targets in cancer. *Nature* **441**: 106–110.

Nicke, B., Bastien, J., Khanna, S.J., *et al.* (2005) Involvement of MINK, a Ste20 family kinase, in Ras oncogene-induced growth arrest in human ovarian surface epithe-lial cells. *Mol Cell* **20**: 673–685.

Nollen, E.A., Garcia, S.M., van Haaften, G., Kim, S., Chavez, A., Morimoto, R.I. and Plasterk, R.H. (2004) Genome-wide RNA interference screen identifies previ-ously undescribed regulators of polyglutamine aggregation. *Proc Natl Acad Sci USA* **101**: 6403–6408.

Nybakken, K., Vokes, S.A., Lin, T.Y., McMahon, A.P. and Perrimon, N. (2005) A genome-wide RNA interference screen in *Drosophila melanogaster* cells for new components of the Hh signaling pathway. *Nat Genet* **37**: 1323–1332.

O'Donnell, K.A., Wentzel, E.A., Zeller, K.I., Dang, C.V. and Mendell, J.T. (2005) c-Myc-regulated microRNAs modulate E2F1 expression. *Nature* **435**: 839–843.

Ovcharenko, D., Jarvis, R., Hunicke-Smith, S., Kelnar, K. and Brown, D. (2005) High-throughput RNAi screening in vitro: from cell lines to primary cells. *Rna* **11**(6): 985–993.

Paddison, P.J., Caudy, A.A., Bernstein, E., Hannon, G.J. and Conklin, D.S. (2002) Short hairpin RNAs (shRNAs) induce sequence-specific silencing in mammalian cells. *Genes Dev* **16**: 948–958.

Paddison, P.J., Silva, J.M., Conklin, D.S., *et al.* (2004) A resource for large-scale RNA-interference-based screens in mammals. *Nature* **428**: 427–431.

Parrish, J.Z., Kim, M.D., Jan, L.Y. and Jan, Y.N. (2006) Genome-wide analyses identify transcription factors required for proper morphogenesis of *Drosophila* sensory neuron dendrites. *Genes Dev* **20**: 820–835.

Pelkmans, L., Fava, E., Grabner, H., Hannus, M., Habermann, B., Krausz, E. and Zerial, M. (2005) Genome-wide analysis of human kinases in clathrin- and caveolae/raft-mediated endocytosis. *Nature* **436**: 78–86.

Philips, J.A., Rubin, E.J. and Perrimon, N. (2005) *Drosophila* RNAi screen reveals CD36 family member required for mycobacterial infection. *Science* **309**: 1251–1253.

Piano, F., Schetter, A.J., Mangone, M., Stein, L. and Kemphues, K.J. (2000) RNAi analysis of genes expressed in the ovary of *Caenorhabditis elegans*. *Curr Biol* **10**: 1619–1622.

Pothof, J., van Haaften, G., Thijssen, K., Kamath, R.S., Fraser, A.G., Ahringer, J., Plasterk, R.H. and Tijsterman, M. (2003) Identification of genes that protect the *C. elegans* genome against mutations by genome-wide RNAi. *Genes Dev* **17**: 443–448.

Poulin, G., Dong, Y., Fraser, A.G., Hopper, N.A. and Ahringer, J. (2005) Chromatin regulation and sumoylation in the inhibition of Ras-induced vulval development in *Caenorhabditis elegans*. *EMBO J* **24**: 2613–2623.

Roos, J., DiGregorio, P.J., Yeromin, A.V., *et al.* (2005) STIM1, an essential and conserved component of store-operated Ca^{2+} channel function. *J Cell Biol* **169**: 435–445.

Rubinson, D.A., Dillon, C.P., Kwiatkowski, A.V., *et al.* (2003) A lentivirus-based system to functionally silence genes in primary mammalian cells, stem cells and transgenic mice by RNA interference. *Nat Genet* **33**: 401–406.

Ryther, R.C., Flynt, A.S., Phillips, J.A., 3rd and Patton, J.G. (2005) siRNA therapeutics: big potential from small RNAs. *Gene Ther* **12**: 5–11.

Schratt, G.M., Tuebing, F., Nigh, E.A., Kane, C.G., Sabatini, M.E., Kiebler, M. and Greenberg, M.E. (2006) A brain-specific microRNA regulates dendritic spine development. *Nature* **439**: 283–289.

Shen, C. and Reske, S.N. (2004) Adenovirus-delivered siRNA. *Methods Mol Biol* **252**: 523–532.

Shirane, D., Sugao, K., Namiki, S., Tanabe, M., Iino, M. and Hirose, K. (2004) Enzymatic production of RNAi libraries from cDNAs. *Nat Genet* **36**: 190–196.

Sieburth, D., Ch'ng, Q., Dybbs, M., *et al.* (2005) Systematic analysis of genes required for synapse structure and function. *Nature* **436**: 510–517.

Silva, J., Chang, K., Hannon, G.J. and Rivas, F.V. (2004a) RNA-interference-based functional genomics in mammalian cells: reverse genetics coming of age. *Oncogene* **23**: 8401–8409.

Silva, J.M., Mizuno, H., Brady, A., Lucito, R. and Hannon, G.J. (2004b) RNA interference microarrays: high-throughput loss-of-function genetics in mammalian cells. *Proc Natl Acad Sci USA* **101**: 6548–6552.

Silva, J.M., Li, M.Z., Chang, K., *et al.* (2005) Second-generation shRNA libraries covering the mouse and human genomes. *Nat Genet* **37**: 1281–1288.

Sonnichsen, B., Koski, L.B., Walsh, A., *et al.* (2005) Full-genome RNAi profiling of early embryogenesis in *Caenorhabditis elegans*. *Nature* **434**: 462–469.

Stegmeier, F., Hu, G., Rickles, R.J., Hannon, G.J. and Elledge, S.J. (2005) A lentiviral microRNA-based system for single-copy polymerase II-regulated RNA interference in mammalian cells. *Proc Natl Acad Sci USA* **102**: 13212–13217.

Stevenson, M. (2004) Therapeutic potential of RNA interference. *N Engl J Med* **351**: 1772–1777.

Szulc, J., Wiznerowicz, M., Sauvain, M.O., Trono, D. and Aebischer, P. (2006) A versatile tool for conditional gene expression and knockdown. *Nat Methods* **3**: 109–116.

Valencia-Sanchez, M.A., Liu, J., Hannon, G.J. and Parker, R. (2006) Control of translation and mRNA degradation by miRNAs and siRNAs. *Genes Dev* **20**: 515–524.

Vastenhouw, N.L., Fischer, S.E., Robert, V.J., Thijssen, K.L., Fraser, A.G., Kamath, R.S., Ahringer, J. and Plasterk, R.H. (2003) A genome-wide screen identifies 27 genes involved in transposon silencing in *C. elegans*. *Curr Biol* **13**: 1311–1316.

Vella, M.C., Choi, E.Y., Lin, S.Y., Reinert, K. and Slack, F.J. (2004) The *C. elegans* microRNA let-7 binds to imperfect let-7 complementary sites from the lin-41 3'UTR. *Genes Dev* **18**: 132–137.

Ventura, A., Meissner, A., Dillon, C.P., McManus, M., Sharp, P.A., Van Parijs, L., Jaenisch, R. and Jacks, T. (2004) Cre-lox-regulated conditional RNA interference from transgenes. *Proc Natl Acad Sci USA* **101**: 10380–10385.

Voorhoeve, P.M. and Agami, R. (2003) The tumor-suppressive functions of the human INK4A locus. *Cancer Cell* **4**: 311–319.

Voorhoeve, P.M., le Sage, C., Schrier, M., *et al.* (2006) A genetic screen implicates miRNA-372 and miRNA-373 as oncogenes in testicular germ cell tumors. *Cell* **124**: 1169–1181.

Westbrook, T.F., Martin, E.S., Schlabach, M.R., *et al.* (2005) A genetic screen for candidate tumor suppressors identifies REST. *Cell* **121**: 837–848.

Yang, F., Zhang, Y., Cao, Y.L., Wang, S.H. and Liu, L. (2005) Establishment and utilization of a tetracycline-controlled inducible RNA interfering system to repress gene expression in chronic myelogenous leukemia cells. *Acta Biochim Biophys Sin (Shanghai)* **37**: 851–856.

Zamore, P.D., Tuschl, T., Sharp, P.A. and Bartel, D.P. (2000) RNAi: double-stranded RNA directs the ATP-dependent cleavage of mRNA at 21 to 23 nucleotide intervals. *Cell* **101**: 25–33.

Zeng, Y. and Cullen, B.R. (2003) Sequence requirements for micro RNA processing and function in human cells. *RNA* **9**: 112–123.

Zhang, S.L., Yeromin, A.V., Zhang, X.H., Yu, Y., Safrina, O., Penna, A., Roos, J., Stauderman, K.A. and Cahalan, M.D. (2006) Genome-wide RNAi screen of Ca^{2+} influx identifies genes that regulate Ca^{2+} release-activated Ca^{2+} channel activity. *Proc Natl Acad Sci USA* **103**: 9357–9362.

Zheng, L., Liu, J., Batalov, S., Zhou, D., Orth, A., Ding, S. and Schultz, P.G. (2004) An approach to genomewide screens of expressed small interfering RNAs in mammalian cells. *Proc Natl Acad Sci USA* **101**: 135–140.

Zhou, H., Xia, X.G. and Xu, Z. (2005) An RNA polymerase II construct synthesizes short-hairpin RNA with a quantitative indicator and mediates highly efficient RNAi. *Nucleic Acids Res* **33**: e62.

High-throughput RNAi in *Caenorhabditis elegans* – from molecular phenotypes to pathway analysis

6

Sarah Jenna and Eric Chevet

6.1 Introduction

Caenorhabditis elegans is a nematode worm comprising approximately 900 cells. Its life cycle is short (about 60 h) and culture conditions are very simple. In addition, cell lineage has been extensively characterized (Brenner, 1974; Sulston, 2003; Sulston and Horvitz, 1977; Sulston *et al.*, 1983). Together, these characteristics made *C. elegans* an organism of choice for genetic and developmental studies (Brenner, 1974). Furthermore, across the years, large collections of mutations in its genome have made this organism one of the most amenable systems for such genetic analyses. In 1998, the *C. elegans* genome was the first metazoan genome to be completely sequenced (http://www.sanger.ac.uk/Projects/C_elegans/) (*C. elegans* Sequencing Consortium, 1998) and this initiative allowed for the generation of resource functional tools such as the *C. elegans* ORFeome (Reboul *et al.*, 2003). These genomic studies were recently utilized to provide the first metazoan protein–protein interaction map (Li *et al.*, 2004).

6.1.1 RNAi in *C. elegans*

RNAi was discovered several years ago following the observation that the introduction of dsRNA caused the specific degradation of mRNA. This was particularly true in *C. elegans* (Guo and Kemphues, 1995) and it was soon recognized as experimentally and technically simple to knock down genes in this organism and other species. Subsequently, it was systematically demonstrated that dsRNA is a potent effector of gene interference (Fire *et al.*, 1998). Although RNAi in *C. elegans* is a powerful method for the inactivation of gene function, it does have several limitations such as tissue- and gene-specific differences in sensitivity to RNAi (Tavernarakis *et al.*, 2000). In addition, it should always be considered that significant inter-experimental variability in RNAi may occur, most likely due to subtle differences in experimental conditions (Simmer *et al.*, 2003).

6.1.2 High-throughput RNAi in *C. elegans*

RNAi technology has allowed a paradigm-shift in the experimental strate-gies developed for the understanding of gene functions at the genome-wide level. A number of groups have subsequently developed methodologies to perform large-scale RNAi in *C. elegans*. Three parameters have to be consid-ered to set up such a screen with (i) the choice of an amenable phenotype to be used as high-throughput readout; (ii) the choice of the dsRNA delivery method; and (iii) the group of genes that will be targeted by the RNAi treat-ment.

Two methods are easily amenable for delivering dsRNA into the *C. elegans* body in a high-throughput fashion: (i) the soaking method, in which worms are immersed in a concentrated dsRNA solution without food for >24 h, and recovered onto conventional culture plates for observation of phenotypes of the soaked worms and their progeny (Maeda *et al.*, 2001; Tabara *et al.*, 1998); and (ii) the feeding method, in which dsRNA-expressing bacteria are fed to worms on agar plates (Timmons and Fire, 1998). A major advantage of the feeding method is that large numbers of worms can be treated at one time, which is of value in biochemical experiments. By allow-ing investigators to select the developmental stage for dsRNA delivery, both the soaking and the feeding methods can be used to conduct stage-specific RNAi experiments.

To perform genome-wide or large-scale RNAi, a DNA library is required to provide the templates for dsRNA synthesis. There are two strategies for the construction of gene libraries for RNAi: use of genomic sequence informa-tion or cDNA libraries. First, genomic PCR products can be amplified based on predicted gene structures to perform functional genomic RNAi analyses. These PCR products can be either *in vitro* transcribed into dsRNA prior to injection (Gonczy *et al.*, 2000) or cloned into a plasmid vector expressing dsRNA in *E. coli* (Fraser *et al.*, 2000; Kamath *et al.*, 2003). Second, cDNA collections can also be used (Maeda *et al.*, 2001; Piano *et al.*, 2000; Reboul *et al.*, 2003). Although PCR product collection based on gene prediction provides a good general representation of the vast majority of genes in the genome, there are also inevitable errors and omissions. In addition, because mature mRNAs are the target molecules in RNAi, misprediction of gene structure might reduce the efficacy of the RNAi response.

6.2 The experiments

Generally, large-scale RNAi analyses may be classified into two categories. The first category is defined as functional genomic analyses aiming to assign gene functions *in vivo* by systematically recording RNAi phenotypes for each gene. The second type of large-scale RNAi involves screening for genes involved in specific processes. In this type of analysis, as in conven-tional forward genetic screens, specific assays are designed to detect abnor-malities in specific phenomena.

In the context of the second type of analysis, we propose to define an experimental setting, in which we can dissect specific signaling pathways leading to the activation of pre-defined transcriptional programs (*Figure 6.1A*). Indeed, in response to specific stimuli, the transcriptional activation of given genes is triggered downstream of specific signaling pathways (Balazsi and Oltvai, 2005; Banerjee and Zhang, 2002; Schlitt and Brazma, 2006). The signaling intermediates, which are transducing the signal (X, X', Y, Y'; *Figure 6.1A*) are leading to the activation of stimulus-dependent inducible genes (i-gene 1 and i-gene 2; *Figure 6.1A*). Conceptually, silencing of these intermediates by RNAi will result in the attenuation of the expression of i-genes (*Figure 6.1A*).

Our first studies were applied to the analysis of a specific stress adaptive response taking place in the endoplasmic reticulum (ER) and named the unfolded protein response (UPR) (Schroder and Kaufman, 2005), which is conserved in *C. elegans* (Shen *et al.*, 2005). The ER is a cellular organelle specialized for folding proteins in transit to the cell surface. Frequently, changes in the extracellular environment result in aberrant protein folding in the ER. The accumulation of improperly folded proteins in the ER leads to adaptive responses, collectively known as the UPR, which induce the expression of genes encoding the protein chaperones and folding catalysts. In this way, the cell up-regulates its protein-folding capacity. A class of novel ER trans-membrane receptors including IRE1, PERK, and ATF6 mediate UPR signal transduction.

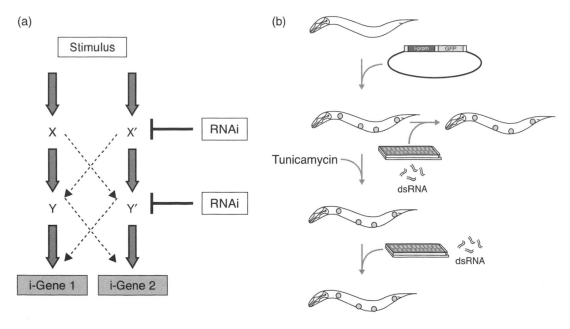

Figure 6.1

Use of transcriptional reporter animals as phenotypic read-out. (A) Basic concept of stimulus-mediated activation of signaling pathways (X, X', Y, Y') which lead to the transcriptional induction of specific genes (i-gene1 and i-gene2; 'i' stands for inducible). RNAi of X, X', Y, Y' is anticipated to prevent X, X', Y, Y'-mediated induction of i-gene1 and/or i-gene2. (B) Schematic representation of our global approach. Transgenic worms expressing GFP under the control of a specific i-gene promoter region are generated and then subjected to RNAi in combination or not with a specific stimulus.

Consequently, our first objective is to create or to use available gene reporter worms expressing GFP under the control of specific UPR-responsive promoters. In the present study and as the prototype experiment for such a screen, we used the *hsp*-4 promoter, which has previously been reported as a target of UPR-dependent pathways through the activation of IRE1 (Urano *et al.*, 2002). Transgenic worms expressing *hsp-4*::GFP were subjected to RNAi treatment in the presence or not of inducers of endoplasmic reticulum stress such as the antibiotic tunicamycin, an inhibitor of *N*-glycosylation (Gu *et al.*, 2004). RNAi was achieved using a feeding procedure with cDNAs derived from the *C. elegans* ORFeome (Protocols 6.1 and 6.2; *Figure 6.1B*).

After a period of time ranging from 4 to 5 days, the average fluorescence emitted by the reporter worms is then quantified using a COPAS (Complex Object Parametric Analyzer and Sorter) Biosort (Union Biometrica). This process is in addition automated and can provide high-throughput quantitative analysis of the fluorescence emitted by a population of worms. Briefly, the COPAS Biosort allows simultaneous excitation and collection of optical measurements of two separate populations. This not only allows us to quantify and average the fluorescence emitted by the worms in response to ER stress induction in combination with RNAi treatment, but also to realize this quantification on an homogenous population of worms (e.g. L1 larvae, adults).

A major strength of our approach resides in the fact that instead of 10–20 worms per condition/RNAi usually tested in high-throughput studies (Fraser *et al.*, 2000; Gonczy *et al.*, 2000; Kamath *et al.*, 2003), our strategy allows for the quantification of hundreds of animals (200 in the protocol described here). As a consequence, this provides a very robust and relevant approach to study transcription regulatory networks applied to ER stress signaling pathways.

Our protocol describes the preparation of cDNA for RNAi studies, the treatment of GFP reporter worms for RNAi by feeding, the quantification of fluorescence using the COPAS Biosort, and finally the analysis and significance of the results.

6.3 Summary

We have applied a high-throughput RNAi approach to study the transcription regulatory networks activated in response to Endoplasmic Reticulum (ER) stress in *C. elegans*. The use of ORF clones from the *C. elegans* ORFeome collection combined with the Gateway® recombinational cloning methodology has dramatically facilitated the RNAi procedure (Protocol 6.1). We propose an RNAi by feeding procedure to selectively silence genes that may be involved in the regulation of ER stress-dependent transcriptional regulation (Protocol 6.2). As previously mentioned, our protocol is applied to the study of ER stress-induced promoters because of the high specificity of this pathway. It is noteworthy that similar protocols can be applied to a large spectrum of promoters downstream of specific inducible signaling pathways such as MAPK, Notch, and FGF.

The combination of RNAi procedures with inducible GFP reporter *C. elegans* strains and quantification of average emitted fluorescence using a COPAS Biosort provides quantitative information on the induction of ER stress. The data can be obtained for specific worm populations (developmental stages, etc.). In addition, complementary qualitative information can also be gathered by microscopic analyses of fluorescent worms. Indeed, this can indicate tissue specificity for the signaling pathways specifically silenced in the RNAi experiments.

In summary, this approach provides an efficient, high-throughput, robust method to analyse specific signaling pathways *in vivo* in a metazoan. We can easily foresee that with the constantly increasing number of GFP-inducible reporter worm strains (http://elegans.bcgsc.ca/home/ge_consortium.html), these experiments may be carried out using multiparallel settings and therefore allow a comprehensive analysis of inducible transcription-regulatory networks in a quantitative manner in living animals.

References

Balazsi, G. and Oltvai, Z.N. (2005) Sensing your surroundings: how transcription-regulatory networks of the cell discern environmental signals. *Sci STKE* **2005**: pe20.

Banerjee, N. and Zhang, M.Q. (2002) Functional genomics as applied to mapping transcription regulatory networks. *Curr Opin Microbiol* **5**: 313–317.

Brenner, S. (1974) The genetics of *Caenorhabditis elegans*. *Genetics* **77**: 71–94.

C. elegans Sequencing Consortium (1998) Genome sequence of the nematode *C. elegans*: a platform for investigating biology. *Science* **282**: 2012–2018.

Fire, A., Xu, S., Montgomery, M.K., Kostas, S.A., Driver, S.E. and Mello, C.C. (1998) Potent and specific genetic interference by double-stranded RNA in *Caenorhabditis elegans*. *Nature* **391**: 806–811.

Fraser, A.G., Kamath, R.S., Zipperlen, P., Martinez-Campos, M., Sohrmann, M. and Ahringer, J. (2000) Functional genomic analysis of *C. elegans* chromosome I by systematic RNA interference. *Nature* **408**: 325–330.

Gonczy, P., Echeverri, C., Oegema, K., *et al.* (2000) Functional genomic analysis of cell division in *C. elegans* using RNAi of genes on chromosome III. *Nature* **408**: 331–336.

Gu, F., Nguyen, D.T., Stuible, M., Dube, N., Tremblay, M.L. and Chevet, E. (2004) Protein-tyrosine phosphatase 1B potentiates IRE1 signaling during endoplasmic reticulum stress. *J Biol Chem* **279**: 49689–49693.

Guo, S. and Kemphues, K.J. (1995) par-1, a gene required for establishing polarity in *C. elegans* embryos, encodes a putative Ser/Thr kinase that is asymmetrically distributed. *Cell* **81**: 611–620.

Kamath, R.S., Fraser, A.G., Dong, Y., *et al.* (2003) Systematic functional analysis of the *Caenorhabditis elegans* genome using RNAi. *Nature* **421**: 231–237.

Lamesch, P., Milstein, S., Hao, T., *et al.* (2004) *C. elegans* ORFeome version 3.1: increasing the coverage of ORFeome resources with improved gene predictions. *Genome Res* **14**: 2064–2069.

Li, S., Armstrong, C.M., Bertin, N., *et al.* (2004) A map of the interactome network of the metazoan *C. elegans*. *Science* **303**: 540–543.

Maeda, I., Kohara, Y., Yamamoto, M. and Sugimoto, A. (2001) Large-scale analysis of gene function in *Caenorhabditis elegans* by high-throughput RNAi. *Curr Biol* **11**: 171–176.

Piano, F., Schetter, A.J., Mangone, M., Stein, L. and Kemphues, K.J. (2000) RNAi analysis of genes expressed in the ovary of *Caenorhabditis elegans*. *Curr Biol* **10**: 1619–1622.

Reboul, J., Vaglio, P., Rual, J.F., *et al.* (2003) *C. elegans* ORFeome version 1.1: experimental verification of the genome annotation and resource for proteome-scale protein expression. *Nat Genet* **34**: 35–41.

Schlitt, T. and Brazma, A. (2006) Modelling in molecular biology: describing transcription regulatory networks at different scales. *Phil Trans R Soc Lond B Biol Sci* **361**: 483–494.

Schroder, M. and Kaufman, R.J. (2005) The mammalian unfolded protein response. *Annu Rev Biochem* **74**: 739–789.

Shen, X., Ellis, R.E., Sakaki, K. and Kaufman, R.J. (2005) Genetic interactions due to constitutive and inducible gene regulation mediated by the unfolded protein response in *C. elegans*. *PLoS Genet* **1**: e37.

Simmer, F., Moorman, C., van der Linden, A.M., Kuijk, E., van den Berghe, P.V., Kamath, R.S., Fraser, A.G., Ahringer, J. and Plasterk, R.H. (2003) Genome-wide RNAi of *C. elegans* using the hypersensitive rrf-3 strain reveals novel gene functions. *PLoS Biol* **1**: E12.

Sulston, J.E. (2003) *Caenorhabditis elegans*: the cell lineage and beyond (Nobel lecture). *Chembiochem* **4**: 688–696.

Sulston, J.E. and Horvitz, H.R. (1977) Post-embryonic cell lineages of the nematode, *Caenorhabditis elegans*. *Dev Biol* **56**: 110–156.

Sulston, J.E., Schierenberg, E., White, J.G. and Thomson, J.N. (1983) The embryonic cell lineage of the nematode *Caenorhabditis elegans*. *Dev Biol* **100**: 64–119.

Tabara, H., Grishok, A. and Mello, C.C. (1998) RNAi in *C. elegans*: soaking in the genome sequence. *Science* **282**: 430–431.

Tavernarakis, N., Wang, S.L., Dorovkov, M., Ryazanov, A. and Driscoll, M. (2000) Heritable and inducible genetic interference by double-stranded RNA encoded by transgenes. *Nat Genet* **24**: 180–183.

Timmons, L. and Fire, A. (1998) Specific interference by ingested dsRNA. *Nature* **395**: 854.

Urano, F., Calfon, M., Yoneda, T., Yun, C., Kiraly, M., Clark, S.G. and Ron, D. (2002) A survival pathway for *Caenorhabditis elegans* with a blocked unfolded protein response. *J Cell Biol* **158**: 639–646.

Protocol 6.1: Generation of constructs driving RNAi through a feeding procedure

Two RNAi-by-feeding libraries are currently available to the scientific community. They have been generated through PCR-amplification of genomic DNA fragments (Fraser *et al.*, 2000; Kamath *et al.*, 2003) or through recombination in RNAi-by-feeding vectors of full-length ORFs of the *C. elegans* ORFeome v1.1 library (Lamesch *et al.*, 2004; Reboul *et al.*, 2003). These two RNAi libraries partially overlap and when combined cover 17 201 genes corresponding to approximately 86% of the predicted worm genome.

For most genes, dsRNA stretches from 200 to 1000 nucleotides (nt) or longer appear to effectively induce interference. However, some specific gene segments are ineffective at inducing interference and it is suggested that dsRNAs from several segments of a gene should be tried. Some additional cloning of cDNA fragments into RNAi-by-feeding vectors is consequently required and could complement existing resources.

Complementary DNA cloning remains a time- and cost-consuming effort. The generation of these constructs using a technology enabling transfer of the coding sequence to various expression vectors constitutes, therefore, a powerful approach to reduce cost related to the identification of genes using RNAi screening and their subsequent functional characterization. The Gateway® cloning technology (Invitrogen) is a high-throughput enabling technology that provides such flexibility. Here, we will detail the different steps required to generate RNAi-by-feeding constructs using this technology and following manufacturer's instructions (http://www.invitrogen.com).

cDNA are synthesized using the Thermoscript RT-PCR system (Invitrogen) after purification of total RNA using RNAzol (Invitrogen). Coding sequences of interest are then PCR amplified using high-fidelity polymerase and specific primers containing attB1 and attB2 recombination sequences. The resulting PCR fragments are inserted into pDONR201 by BP recombination. BP products are then transformed into *E. coli* DH5-alpha strain and plated in six-well LB-agar plates containing 100 µg/ml ampicillin. After an overnight incubation at 37°C, the clones isolated are screened by PCR for the presence of the coding sequence in pDONR201 (*Table 6.1*). Plasmids are purified and cDNA fragments cloned in the Entry vector sequenced.

The resulting open reading frames (ORFs) or cDNA fragments are then transferred from pDONR201 into pL4440-dest-RNAi, a Gateway®-compatible RNAi vector adapted from the original pL4440-RNAi vector (25). For fragments larger that 1000 bp it is recommended to digest pL4440-dest-RNAi vector with EcoRI and NcoI before the LR reaction. This will significantly improve the efficiency of the LR reaction. LR reactions are performed as indicated in the Gateway® cloning manual.

The efficiency of the RNAi-by-feeding procedure requires that bacteria could express and accumulate a large amount of dsRNA. For this reason RNAi mediating constructs are transformed

Table 6.1 Oligonucleotides to be used for PCR screens

PCR screen on pDONR201	PCR screen on pL4440-dest
pDONR201-F: 5'CGCGTTAACGCTAGCATGGATCTC pDONR201-R: 5'GTAACATCAGAGATTTTGAGACAC	pL4440-dest-RNAi-F: 5'GTTTTCCCAGTCACGACGTT pL4440-dest-RNAi-R: 5'TGGATAACCGTATTACCGCC

in an engineered *E. coli* HT115 (DE3) strain, which is RNAse III-deficient and able to express T7 polymerase upon IPTG induction. LR products could not be directly transformed into HT115 (DE3) competent bacteria since this RNase III-deficient strain of *E. coli* is resistant to the *ccd*B toxic gene used to select recombined pL4440 clones. Therefore, LR products are transformed into *E. coli* DH5-alpha strain and plated in six-well LB-agar plates containing 100 µg/ml ampicillin. After an overnight incubation, the clones isolated are screened for the presence of the pL4440-ORF using PCR (*Table 6.1*). Plasmid DNA minipreps are then prepared and DNA preparations subsequently transformed into HT115 (DE3) bacteria. The genotype of this strain is F-, *mcrA, mcrB, IN(rrnD-rrnE)1, lambda-, rnc14::Tn10* (DE3 lysogen: *lacUV5* promoter-T7 polymerase). T7 polymerase gene expression is driven by the *lacUV5* promoter that is IPTG inducible. *Rnc14* encodes RNAse III that is disrupted by Tn10 and consequently unable to degrade the dsRNA expressed *in vivo*. Tn10 carries a tetracycline-resistant gene. The transformed bacteria are then subjected to two antibiotic selections 100 µg/ml ampicillin and 12.5 µg/ml tetracycline.

A pool of eight colonies for each construct are grown in LB medium containing 100 µg/ml ampicillin and 12.5 µg/ml tetracycline overnight at 37°C. Sterile glycerol is added to the culture to reach an 8% final concentration. The bacterial stocks can be stored at −80°C for periods up to 2 years.

Protocol 6.2: RNAi treatment of GFP reporter animals

Transgenic *C. elegans* strains expressing *hsp-4*::GFP have been used in several studies as ER-stress reporters (Urano *et al.*, 2002). The BC11945, *dpy-5(e907)*, sEX11945-[*dpy-5(+)* + rCes-Cb-*hsp-4*-GFP+pCes361] transgenic strain generated by the *C. elegans* gene expression consortium (http://elegans.bcgsc.ca/home/ge_consortium.html) is used in our study. These animals express GFP reporter under the control of *hsp-4* promoter sequence.

Some mutations have been demonstrated to increase sensitivity to dsRNA treatment: *eri-1(mg366)* IV and *rrf-3(pk1426)* II (19). RNAi phenotypes in *rrf-3(pk1426)* II genetic background may be stronger and more closely resemble a null phenotype as compared with wild-type phenotype (Simmer *et al.*, 2003). However, this mutation was shown to silence transgenes and is therefore not recommended to examine GFP expression in transgenic animals. The generation of transgenic animals expressing *hsp-4*::GFP in the genetic background of *eri-1(mg366)* IV mutation could constitute a excellent model to increase RNAi-induced phenotype expressivity and to investigate ER stress mechanisms in neurons that are resistant to RNAi treatment in wild-type strains. In this study, we decided to identify UPR regulators in somatic and non-neuronal cells and consequently performed our screening using wild-type animals carrying *hsp-4*::GFP transgene as an extrachromosomal array.

To date, RNAi screens using *hsp-4*::GFP expressing animals as UPR reporter remained low-throughput, the bottle neck of such studies consisting in the recording of GFP expression through visual inspection using fluorescence microscopy. The recent development and commercialization of the COPAS Biosort by Union Biometrica now provide the necessary technology for the development of high-throughput RNAi screening for UPR regulators. The COPAS Biosort enables automated quantitative analyses of GFP emission in living animals, but also sorting and dispensing of worm populations. To automate the analysis of RNAi treated worm populations, the COPAS Biosort is interfaced with a ReFLx sampler module. This COPAS module has been designed to facilitate analysis of populations incubated in 96-well plates. The ReFLx probe aspirates the worm population from the well, washes it, filters it and reroutes it through the flow cell for analysis or for sorting. The ReFLx unit can analyze and sort a 96-well plate in less than an hour with a yield superior to 90%. This equipment is used in our screen to sort GFP expressing animals and to measure in a quantitative and automated manner GFP emission of animals submitted to RNAi and drug treatments.

Our approach consists of the identification of genes required for tunicamycin-induced *hsp-4* induction but also of those whose expression is required for the maintenance of ER homeostasis. To this end, worms are subjected to RNAi treatment and the UPR measured under basal conditions or after incubation with tunicamycin. Tunicamycin is an antibiotic inhibitor of *N*-glycosylation and was used as an ER stressor in numerous animal models. Each treatment combination is performed in triplicate, and each plate contains the five following controls: (i) worms fed with HT115 (DE3) bacteria transformed with empty pL4440 or (ii) pL4440-GFP vector represent negative and positive RNAi controls, respectively. RNAi treatment with pL4440-GFP is used to quantify GFP emission background while feeding with bacteria transformed with empty vector will enable the measurement of UPR (as reported by *hsp-4* induction) in wild-type animals. RNAi treatments against *ire-1*, *pek-1* and *atf-6* are present in every RNAi series/plate and are used to normalize results on a reference experiment. Using this protocol, 160 novel genes can be tested for their potential regulatory function on UPR per month and per person.

DAY 1

As shown in *Figure 6.2*, 1 ml of LB medium containing 100 µg/ml ampicillin is inoculated with HT115 clones and grown for 8 h at 37°C upon agitation in a 2-ml squared-bottom 96-well block sealed with a breathing plate mat. dsRNA synthesis is induced in *d1* cultures by addition of 1 mM IPTG and 16 h incubation at 22°C. In the mean time, 10 × 100-mm NGM plates containing a large

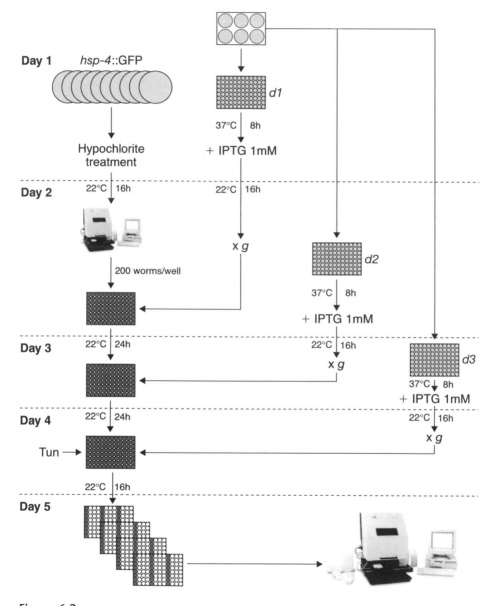

Figure 6.2

RNAi-by-feeding procedure. The whole procedure is detailed in the text (Protocol 6.2). Gray circles represent 100-mm diameter NGM plates containing adult hermaphrodites. Light gray shaded wells represent bacterial liquid cultures and dark gray shaded wells represent NGM-agar seeded with bacteria and worms. '× *g*' stands for centrifugation at 3000 × *g* for 20 min. Tun, treatment with 5 µg/ml tunicamycin.

amount of adult hermaphrodites are submitted to sodium hypochlorite treatment. Briefly, worms are floated from the plate using 10 ml of M9 buffer and incubated for 5 min in 0.4 M hypochlorite buffer (*Table 6.2*). Eggs are subsequently washed five times and resuspended in 20 ml of sterile M9 buffer. Eggs are incubated 16 h at 22°C upon agitation. During this incubation, eggs will hatch and *C. elegans* development will arrest in L1 due to the absence of food.

Table 6.2 M9 and hypochlorite buffers

M9 Buffer 1X	0.5 M hypochlorite buffer
5.8 g Na_2HPO_4	10 ml Na–hypochlorite 4% stock solution
3 g KH_2PO_4	15 ml dH_2O
0.5 g NaCl	25 ml NaOH 1 M stock solution
1 g NH_4Cl	
Bring to 1 l with dH_2O and autoclave	

DAY 2

D1 cultures are centrifuged at 3000 × *g* for 20 min. LB culture medium is removed from the wells and the bacterial pellet resuspended with 10 μl of LB medium. Flat-bottom 96-well plates containing 100 μl of NGM-agar containing 100 μg/ml ampicillin (*Table 6.3*) are seeded with *d1* cultures. Agar wells are air-dried in a sterile environment for up to 30 min. COPAS Biosort is used to sort GFP expressing L1 Larvae from the population synchronized in day 1, and to dispense 200 animals per well (Protocol 6.3). Agar wells are air-dried for 1 h and subsequently incubated for 24 h in a humid chamber at 22°C. HT115 *d2* cultures are inoculated and induced for dsRNA expression as done for *d1* cultures.

Table 6.3 NGM-agar plate

NGM classic
12 g NaCl
64 g Agar
10 g Bacto-peptone
4 L dH_2O
Autoclave and add:
4 ml cholesterol 5 mg/ml in 95% EtOH stock solution
4 ml $MgSO_4$ 1 M stock solution
4 ml $CaCl_2$ 1 M stock solution
4 ml KPO_4 pH 6.0, 1 M stock solution

DAY 3

D2 cultures are centrifuged and used to feed the worms as done in day 2 for *d1* cultures. Agar plates are air-dried for up to 30 min and incubated for an additional 24 h in a humid chamber at 22°C. *D3* cultures are inoculated and induced as done for cultures *d1* and *d2*.

DAY 4

Worms are fed with *d3* cultures and ER stress induced by adjunction of 5 μg/ml of tunicamycin to the agar plates. Plates are air-dried and incubated for 16 h at 22°C.

DAY 5

Worms are floated off the plate in M9 buffer (*Table 6.1*) containing 0.01% Triton X-100, and dispensed in conical 96-well plates as shown in *Figure 6.2*. Three wells filled with COPAS sheath buffer are placed between wells containing worms. This is required to wash the COPAS tubules system and therefore to avoid cross-contamination between wells.

Protocol 6.3: Sorting of fluorescent animals and measurement of the UPR

Sorting of L1 animals expressing GFP as well as measurement of GFP expression after RNAi and drug treatment is carried out using the Union Biometrica COPAS Biosort. The COPAS Biosort is equipped with two lasers. A red diode excitation laser is used to analyze the physical parameters of the organism, referred to as time of flight (TOF) and extinction (EXT). TOF is a measure of the relative length of each organism, and EXT provides measurement of its optical density. A multiline argon laser is used to excite various fluorophores. The 499-nm line of the laser is used to excite the green fluorescent reporter. The design of the COPAS allows simultaneous excitation and collection of optical measurements of two separate populations.

SORTING OF L1 EXPRESSING GFP

Worm populations are synchronized in L1 after hypochlorite treatment and 16 h incubation with sterile M9. Almost 50% of *hsp-4*::GFP expressing animals lose their transgene at each generation. The percentage of animals still able to express GFP being variable from one generation to another, the animals that have lost the array have to be eliminated from the population that will be submitted to RNAi treatment. To this end, the population at an approximate concentration of one animal per microliter is placed in the COPAS cup. After running a small portion of the sample through the Biosort, two parameters the TOF and the EXT are used to analyze the population. A gating region (R1) is drawn on an EXT versus TOF dot-plot to eliminate dead eggs and debris (*Figure 6.3A*). The sorting dot plot is then set so that the FLU1 (GFP signal) and TOF parameters are displayed. A

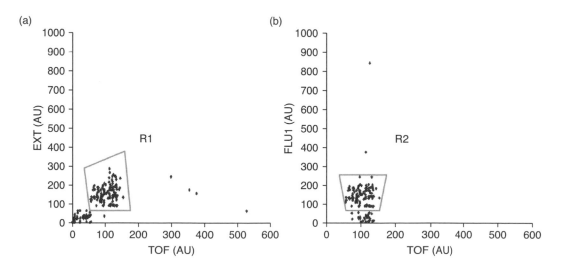

Figure 6.3

Sorting of GFP-expressing L1 larvae. (A) Identification of L1 animals based on EXT and TOF parameters. (B) Sorting of GFP-expressing animals based on TOF and FLU1 parameters.

sorting region (R2), selecting the L1 animals expressing the GFP, is then drawn on the FLU1 versus TOF dot plot (*Figure 6.3B*). Two hundred worms are sorted and dispensed per well of the NGM-agar 96-well plate.

QUANTIFICATION OF *HSP-4*::GFP EXPRESSION

Populations submitted to RNAi and treated or not by tunicamycin are processed through the COPAS using the ReFLx module. The quantifications provided by the COPAS are retrieved and analyzed. To identify adult worms from L1 and debris, worms are sorted on the basis of their TOF and green fluorescence emission (FLU1). The TOF measures the relative length of each animal by measuring pulse width. A count of 100 corresponds to an individual of approximately 0.24 mm long. Adults are 1 mm long, and are consequently characterized by a TOF of between 400 and 500. GFP emission (FLU1) of the adults detected in each adult population is normalized to the GFP emission of adults detected in population fed with HT115 bacteria transformed with pL4440-GFP construct. After normalization, the average GFP emission and the mean value are calculated for each worm population.

Average GFP emissions from animals treated by RNAi against *ire-1*, *atf-6* and *xbp-1* are compared with a reference experiment. This comparison enables the calculation of a variation factor β. This variation factor is used to normalize results obtained for each RNAi series to a reference experiment. Consequently, this helps to reduce the fluctuation of GFP expression resulting from environmental variability.

Comparison of GFP expression measured in basal conditions and upon tunicamycin treatment enables the identification of at least three classes of genes: (i) the genes whose silencing does not affect *hsp-4*::GFP expression; (ii) the genes whose silencing prevents tunicamycin-induced *hsp-4*::GFP expression (activators); (iii) and the genes whose silencing leads to increased basal *hsp-4*::GFP expression (repressors). Activators encode proteins whose functional alteration would have a direct effect on ER homeostasis. In opposition, repressors encode products that are directly involved in the mediation of the UPR signaling pathways towards the *hsp-4* promoter.

RNAi in *Xenopus laevis*

Adrianna L. Stromme and Craig A. Mandato

7

7.1 Introduction

The South African clawed frog, *Xenopus laevis*, has been used as a model system for the research of developmental biology for many decades. Undoubtedly, the large size of the amphibian oocytes makes for easy manipulation of individual cells. Upon fertilization, rapid development of these large (~1 mm diameter) cells allows for easy visualization of cleavage events and vertebrate development. Furthermore, these animals require low maintenance, are inexpensive and easy to care for, and most importantly give rise to large numbers of oocytes and eggs. While using the *X. laevis* model system clearly has its advantages, there are shortcomings, notably the animal's long breeding cycle and tetraploidy. In fact, these disadvantages have in the past made loss-of-function studies such as gene knockouts difficult and complicated. With the discovery that the *X. laevis* oocyte can translate microinjected mRNA (Gurdon and Lane, 1971), came the advent of overexpression studies, and early forms of loss-of-function methods such as the use of antisense RNA or dominant negative mRNA. These techniques, however, have their caveats as well; a high concentration of antisense RNA is necessary for this approach to work accordingly and the outcome is often transient (Dirks *et al.*, 2003) and non-specific (Zhou *et al.*, 2002).

The discovery of RNAi and the finding that dsRNA can prevent gene expression has been a major breakthrough for loss-of-function studies in many different organisms. RNAi was characterized in *X. laevis* and other mammals much later than such organisms as plants and *Drosophila melanogaster,* due to early accounts of long dsRNA causing non-specific results in vertebrate model systems (Brummelkamp *et al.*, 2002; Zhou *et al.*, 2002). A cytotoxic interferon response is activated upon addition of dsRNA longer than 30 nt, which subsequently gives rise to the degradation of mRNA in a non-specific manner and ultimately cell death in many cases (Brummelkamp *et al.*, 2002; Dirks *et al.*, 2003; Elbashir *et al.*, 2001). However, this problem was eliminated with the discovery that siRNA of approximately 21–23 nt can actually avoid the non-specific results obtained from using long dsRNA (Elbashir *et al.*, 2001). This recent finding has not only lead to the use of siRNA in vertebrate organisms, but allowed for a fast and relatively inexpensive technique to be used for loss-of-function studies in animals such as *X. laevis* where earlier approaches were problematic.

The first reported use of RNAi on *X. laevis* embryos came from a study on *Xlim-1*, a LIM class homeobox gene located in the Spemann organizer, which was thought to play a role in stimulation and induction of neural tissue (Nakano *et al.*, 2000). To first determine if RNAi was an effective technique, an exogenous reporter gene expressing luciferase was coinjected into two blastomeres at the four-cell stage alongside dsRNA for the luciferase gene.

Indeed, compared with the injection of double-stranded β-globulin (negative control) or luciferase alone, the injection of the double-stranded luciferase RNA caused a very significant decrease in luciferase activity (Nakano *et al.*, 2000). Since it appeared that RNAi was working for an exogenous gene such that the dsRNA was interrupting the expression of the luciferase gene, dsRNA of *Xlim-1*, an endogenous gene was then tested. Previous findings revealed that in mice, *Lim-1* knockouts resulted in animals that lacked normal head structures. Therefore, it was postulated by Nakano *et al.* that the use of ds*Xlim-1* RNA to interrupt the *Xlim-1* gene in *Xenopus* embryos would cause similar deficits in head development. Both dsRNA encoding *Xlim-1* and nβ-gal mRNA as a lineage tracer, were microinjected into the two blastomeres at the four-cell stage of the embryo. As predicted, by down-regulating the *Xlim-1* gene, head defects were visualized and subsequent RT-PCR confirmed that endogenous levels of *Xlim-1* mRNA were diminished with the use of ds*Xlim-1* (Nakano *et al.*, 2000). Taken together, the results of this early study using RNAi on the *X. laevis* system undoubtedly paved the way for future loss-of-function research to be done on vertebrate models.

Even though Nakano *et al.* had established that RNAi was possible via dsRNA, there were unresolved concerns regarding the non-specific effects of using long dsRNA. The use of siRNA in *Xenopus* had not yet been studied and the mode of action of RNAi via siRNA, namely whether or not RNAi occurred transcriptionally or post-transcriptionally in vertebrate systems, was a question that remained outstanding. The first report of siRNA use in the *X. laevis* model system established that siRNA functions to suppress genes in a post-transcriptional manner (Zhou *et al.*, 2002). Initially an exogenous reporter gene (luciferase) was injected alongside siRNA targeting luciferase and the expression of the reporter gene was decreased. Furthermore, Zhou *et al.* determined that siRNA gene inhibition is sequence dependent; microinjecting siRNA targeting luciferase that is different by 3 nt resulted in less of a decrease in gene expression. siRNA was then directed against endogenous genes (cyclin B1 and B2) and there was a significant reduction in expression at the 32 cell stage. It was subsequently determined that because embryos do not undergo transcription prior to mid-blastula transition (MBT), the observed lack of expression at the 32-cell stage is due to post-transcriptional inactivation of mRNA (Zhou *et al.*, 2002). To test this theory, luciferase mRNA was injected into two-cell stage embryos along with the corresponding luciferase encoding siRNA and again a decrease in luciferase expression was visualized after MBT. This thorough study provided evidence for the ability of siRNA to work appropriately in the *X. laevis* model and that its mechanism in vertebrates is most likely via post-transcriptional eradication of intended mRNA.

Since the time at which these two major studies were published, siRNA has been used as a valuable tool to knock down various proteins involved in a range of subjects from ion channels (Anantharam *et al.*, 2003; Gordon *et al.*, 2006), to synaptic receptor subunits (Miskevich *et al.*, 2006), and even hormone receptors (Haas *et al.*, 2005). It should be mentioned that there are limitations to using siRNA in *X. laevis*, such as the difficulty in silencing genes at later stages of development (Li and Rohrer, 2006). Moreover, the future of gene silencing in *Xenopus* may lie in using a combination of transgenesis and RNAi (Dirks *et al.*, 2003; Li and Rohrer, 2006), but for now this method of siRNA alone is fast, inexpensive, and accurate.

7.2 Oocyte isolation

7.2.1 Inducing ovulation

The number and quality of eggs released from female frogs can be extremely variable. Therefore, when collecting unfertilized eggs for experimental purposes, it is wise to actually induce ovulation in at least two frogs to guarantee collection of viable cells. It should also be noted that frogs should not be fed prior to inducing ovulation. Human chorionic gonadotropin (hCG) is used to induce ovulation, and it is injected into the dorsal lymph sac. Purchased hCG (ICN Biomedicals, Costa Mesa, CA) should be resuspended in sterile water (2000 units/ml) and using a 26-gauge needle, 500–900 U should be injected into the posterior and medial aspect of the frog known as the dorsal lymph sac (Mandato *et al.*, 2000; Sive *et al.*, 1997; Zhou *et al.*, 2002). The needle should be inserted under the skin of the frog, closer to the hind limbs and slightly away from the visible lateral lines. Moving the needle towards the dorsal midline and subsequently across the lateral lines, the dorsal lymph sac can be located between lateral lines and at the back, posterior area of the frog. The wall of the sac can be felt if the needle is inserted correctly and after injecting the hCG one should wait a few seconds before removing the needle. No blood or hCG should seep out of the needle wound if this is done properly (Sive *et al.*, 1997).

On the other hand, if frogs are new or induction of ovulation has not been performed on the animal for over a year, female frogs are often primed with injections of hCG 5 days before normal induction of ovulation (Sive *et al.*, 1997). Furthermore, priming is also conducted during certain times of the year when the frogs may not be ovulating well due to seasonal changes; this is common during the months of August and/or December. It should be noted that priming *does not* improve the quality of the egg, but increases the quantity. Injections of approximately 50 U of hCG are given 5 days prior to scheduled induction; higher doses are avoided otherwise fewer eggs may be laid (Sive *et al.*, 1997).

Approximately 8–10 h after inducing ovulation, the frog should commence laying eggs; if kept in cooler temperatures (15–17°C) this procedure may take longer (Sive *et al.*, 1997). Frogs should be allowed to recover for at least 8 weeks before inducing ovulation again and if possible, 4 months is even better.

7.2.2 Collecting eggs

It is of utmost importance that the water in which the female frogs are kept during and after induction of ovulation is kept fresh and clean. The frogs are vulnerable to septic shock during this period, and some labs even prefer to keep the females in a salt and antibiotic added environment to ensure frogs are kept healthy (Sive *et al.*, 1997).

Collection of unfertilized eggs is done physically and meant to imitate the male frog during normal fertilization events. Techniques vary, but the frog should be held by the thigh and massaged on its belly using lateral strokes and constant, gentle pressure (Mandato *et al.*, 2000). Soon after this, the frog will begin to lay eggs, which should be captured by squeezing and holding the animal over a petri dish. A solution of MBS and salts (Protocol 7.1) should be added to the acquired eggs. Eggs can be collected approximately every hour, but the frog should not be physically massaged, as described above, for longer than 2–3 min (Mandato *et al.*, 2000).

7.3 Testes isolation

The testes are isolated in order to conduct *in vitro* fertilization and the male frogs can either be killed or anesthetized in benzocaine (Protocol 7.1) to perform this surgery (Mandato *et al.*, 2000). A fully mature male frog should be chosen so as to obtain optimal fertilization rates; a mature male frog can be discerned from an immature frog via the large size of the nuptial pads on its forelimbs. The anesthetized frog should be laid on its back, and an incision should be made on its lower abdomen. Cutting through the visceral layer of the abdominal cavity, and pushing aside the liver should help to reveal fat bodies (yellow in color) on which the testes lie at the base (Sive *et al.*, 1997). The testes appear whitish and are encompassed by vasculature and can be removed via scissors (Sive *et al.*, 1997). Surgery done under a dissection microscope can also facilitate the procedure, although it is not necessary. Upon removal, the testes should be placed in a solution of serum and Marc's Modified Ringer solution (MMR) until ready for use (Protocol 7.1).

7.4 *In vitro* fertilization

Fertilization is optimal if performed as soon as possible after collection of the eggs and after isolation of the testes. Techniques vary at this point, but the main idea is that every egg should come in contact with a portion of the testes. This can be done by manually touching the eggs with a piece of cut up testis or the testes can be minced and this mixture can be added to the petri dish housing the eggs (Mandato *et al.*, 2000; Sive *et al.*, 1997; Zhou *et al.*, 2002). Prior to adding the crushed testes, most of the buffer solution should be removed from the petri dish of eggs, this allows for the sperm to have easier access to the eggs (Sive *et al.*, 1997). The binding of sperm to the egg is instantaneous so there is no need for a long incubation; therefore, shortly after the addition of the testis, the petri dish of eggs should be washed with a dilute salt solution of MMR (Protocol 7.1).

Upon fertilization, the pigmented animal hemisphere contracts upwards such that the majority of the egg is white or unpigmented in color. To prevent polyspermy, the egg also becomes stiffer, which is another sign that the fertilization was successful. Furthermore, fertilized eggs rotate after approximately half an hour and cause the animal pole to always face upwards (Sive *et al.*, 1997).

7.5 Microinjecting dsRNA into embryos/oocytes

7.5.1 Dejellying embryos

A thick, sticky jelly coat surrounds the *X. laevis* embryo that serves to protect and help with its development. In order for any manipulation to be performed on the embryo, this jelly coat must first be removed. The embryos can be dejellied by submerging and gently swirling them in 2–3% cysteine in MMR (Protocol 7.1); it is of utmost importance that the pH be tested frequently and be held around pH 7.8–8 (Mandato *et al.*, 2000). This swirling should be done for approximately 4 min but can vary. The jelly can be visualized floating to the top of the dish and at this point the embryos can again be rinsed repeatedly in a 0.1 M solution of MMR (Mandato *et al.*, 2000; Sive *et al.*, 1997). An alternative technique for dejellying embryos is to treat them in a solution of DTT and HEPES (Protocol 7.1). This solution should be left on for roughly 4 min and no longer, as the DTT is toxic and may damage the embryos after their coats have come off. Once again, with this technique the embryos should be rinsed many times afterwards in a 0.1 M solution of MMR (Mandato *et al.*, 2000).

7.5.2 Vitelline membrane removal

The main role of the vitelline envelope is to prevent polyspermy; when a sperm fuses with the egg, calcium is released into the perivitelline space and subsequent hardening of the vitelline membrane ensues. This membrane, however, becomes a nuisance when it comes to micromanipulating the embryo and removal facilitates the entire process. A pair of forceps is needed to dispose of this membrane manually, as one pair should be blunt in order to grasp the embryo, and the other set should be sharp and pointy to allow for easy removal of the vitelline envelope (Sive *et al.*, 1997). Proteinase K (5 µg/ml) can also be used if the membranes are stubborn and not loosening easily (Sive *et al.*, 1997). Manual removal, however, is more desirable since treating the embryos with enzymes can cause damage.

7.5.3 Microinjections

While amounts vary, anywhere from 500 pg to 10 nl of double-stranded siRNA has been reported as successfully injected into embryos (Anantharam *et al.*, 2003; Haas *et al.*, 2005; Miskevich *et al.*, 2006). In order to calibrate the amount of siRNA to microinject, a micromanipulator apparatus should be set up with an attached capillary tube needle. Using the pressure from the micromanipulator, water can be aspirated up the capillary tube. A hemacytometer should next be set up with a drop of oil on its grid. To calculate the amount needed to inject, the following calculation can be used.

Amount needed = $4/3 \times \pi \times r^3$

The maximum amount of additional fluid that can be added to an oocyte is approximately 40 nl, while 5–10 nl should be the maximum amount added to a one-cell embryo. For siRNA injections, the amount to inject varies depending on which stage the embryo is in; for a one-cell embryo, 10 nl works well (Miskevich *et al.*, 2006). Knowing the size of the grid lines on the hemacytometer and using the above equation, the diameter of a sphere can be calculated. By injecting the water-filled capillary needle and injecting some onto the oil-covered hemacytometer, the diameter of a water bubble can be adjusted to the appropriate size, thereby calibrating the amount of siRNA needed to be injected. Once the micromanipulator is set to the appropriate pressure to obtain the proper volume ejected, embryos can be microinjected accordingly. Injections can be done from one-cell stage onwards but it must be remembered that the amount injected into cells at further cleavages should be decreased accordingly since the cells are getting smaller in size. Care should be taken with the needle bore size; if it is too big, irreversible damage can occur and cytoplasm will leak out of the large hole (Sive *et al.*, 1997).

7.6 Lineage labeling

Injected into embryos to target them at later time intervals, tracer molecules are inert and allow for the visualization of cells that have been manipulated (Sive *et al.*, 1997). Here are three examples of lineage tracers that have been found to work well in the *X. laevis* model system.

7.6.1 Dextran amines

Dextran amines diffuse rapidly over short distances and have been used extensively for neuronal tract tracing, but these molecules can be tagged with fluorescent labels such as rhodamine (RDA) and used for lineage labeling as well (Fritzsch, 1993; Sive *et al.*, 1997). Not only are they inert, but are sustained throughout fixation and can also be used for live cell imaging. However, there are some accounts that dextran amines lead to less translation (Sive *et al.*, 1997) and are problematic to visualize, since particular filters must be used to do so. If using fluorescently conjugated dextran amines, they should be resuspended in phosphate-buffered saline (PBS) buffer (Protocol 7.1) at 15–120 mg/ml and the maximum volume that should be injected for one-cell stage cells is approximately 50 nl (Sive *et al.*, 1997).

7.6.2 β-Galactosidase RNA

A common technique for lineage labeling is to co-inject an RNA tracer alongside the RNA of interest. This is advantageous because the co-injection will allow for both RNAs to diffuse throughout the cell at the same pace (Sive *et al.*, 1997). β-Galactosidase in conjunction with a nuclear localization signal is the most frequently used RNA tracer, as X-gal staining can be used to visualize the β-gal RNA in the embryo (Nakano *et al.*, 2000; Sive *et al.*, 1997). β-Galactosidase is also beneficial due to its ability to label cells for long periods of time; its stability allows for embryos to be followed well into the tadpole stages (Nakano *et al.*, 2000; Sive *et al.*, 1997). The major caveat to using β-galactosidase is that fixation is necessary to detect staining, so therefore, if live cell imaging is needed, this protocol should not be used (Protocol 7.2).

7.6.3 GFP RNA as a lineage marker

For tracing lineage in living cells, the most popular technique is to use injections of GFP RNA. After microinjecting, the cells can be visualized for a short window of time (approximately 4–8 h), after which the GFP signal begins to dissipate (Sive *et al.*, 1997).

7.7 Screening of phenotypes

Phenotypes are differentiated by visualization under a light microscope. Generally, when looking at embryos and even later stages such as tadpoles, a 10× objective is ample for discerning between phenotypes. However, when specimens are injected with dextran amines or GFP RNA, an epifluorescent microscope with the appropriate filters should be used to visualize the fluorescence from these lineage labeling techniques.

References

Anantharam, A., Lewis, A., Panaghie, G., Gordon, E., McCrossan, Z.A., Lerner, D.J. and Abbott, G.W. (2003) RNA interference reveals that endogenous *Xenopus* MinK-related peptides govern mammalian K+ channel function in oocyte expression studies. *J Biol Chem* **278**: 11739–11745.

Brummelkamp, T. R., Bernards, R. and Agami, R. (2002) A system for stable expression of short interfering RNAs in mammalian cells. *Science* **296**: 550–553.

Dirks, R., Bouw, G., Huizen, R.V., Jansen, E.J.R. and Martens, G.J.M. (2003) Functional genomics in *Xenopus laevis*: towards transgene-driven RNA interference and cell-specific transgene expression. *Curr Genomics* **4**: 699–711.

Elbashir, S.M., Harborth, J., Lendeckel, W., Yalcin, A., Weber, K. and Tuschl, T. (2001) Duplexes of 21-nucleotide RNAs mediate RNA interference in cultured mammalian cells. *Nature* **411**: 494–498.

Fritzsch, B. (1993) Fast axonal diffusion of 3000 molecular weight dextran amines. *J Neurosci Methods* **50**: 95–103.

Gordon, E., Roepke, T.K. and Abbott, G.W. (2006) Endogenous KCNE subunits govern Kv2.1 K+ channel activation kinetics in *Xenopus* oocyte studies. *Biophys J* **90**: 1223–1231.

Gurdon, J.B. and Lane, C.D. (1971) Use of frog eggs and oocytes for the study of messenger RNA and its translation in living cells. *Nature* **233**: 177–182.

Haas, D., White, S.N., Lutz, L.B., Rasar, M. and Hammes, S.R. (2005) The modulator of nongenomic actions of the estrogen receptor (MNAR) regulates transcription-independent androgen receptor-mediated signaling: evidence that MNAR participates in G protein-regulated meiosis in *Xenopus laevis* oocytes. *Mol Endocrinol* **19**: 2035–2046.

Li, M. and Rohrer, B. (2006) Gene silencing in *Xenopus laevis* by DNA vector-based RNA interference and transgenesis. *Cell Res* **16**: 99–105.

Mandato, C.A., Weber, K.L., Zandy, A.J., Keating, T.J. and Bement, W.M. (2000) *Xenopus* egg extracts as a model system for analysis of microtubule, actin filament and intermediate filament interactions. In: *Cytoskeletal Methods and Protocols* (ed. R.H. Gavin). Humana Press Inc., Totowa, NJ, pp. 229–239.

Miskevich, F., Doench, J.G., Townsend, M.T., Sharp, P.A. and Constatine-Paton, M. (2006) RNA interference of *Xenopus* NMDAR NR1 *in vitro* and *in vivo*. *J Neurosci Methods* **152**: 65–73.

Nakano, H., Amemiya, S., Shiokawa, K. and Taira, M. (2000) RNA interference for the organizer-specific gene Xlim-1 in *Xenopus* embryos. *Biochem Biophys Res Commun* **274**: 434–439.

Sive, H., Grainger, R. and Harland, R. (1997). *Early development of* Xenopus laevis *course manual, Cold Springs Harbour*. Cold Springs Harbour Lab Press, Cold Springs Harbour, New York.

Zhou, Y., Ching, Y., Kok, K.H., Kung, H. and Jin, D. (2002) Post-transcriptional suppression of gene expression in *Xenopus* embryos by small interfering RNA. *Nucleic Acids Res* **30**: 1664–1669.

Protocol 7.1: Solutions appendix

MMR (MARC'S MODIFIED RINGER SOLUTION)

1000 mM NaCl
20 mM KCl
10 mM $MgCl_2$
20 mM $CaCl_2$
50 mM HEPES
= 10 × stock solution, pH 7.5, autoclave before using, dilute accordingly. After fertilization, use 1/3–1/10 dilutions of MMR to rinse and wash embryos

MBS (MODIFIED BARTH'S SOLUTION) – 1 ×

88 mM NaCl
5 mM HEPES
2.5 mM $NaHCO_3$
0.7 mM $CaCl_2$
1 mM $MgSO_4$
1 mM KCl

The pH should be maintained at 7.5; for use after egg collection, a high salt 1 l solution should be made of the following: 7 ml of 0.1 M $CaCl_2$, 100 ml of 10 × MBS, 4 ml of 5 M NaCl and diluted to 1 l with water (Sive *et al.*, 1997).

FROG ANESTHETIC (BENZOCAINE)

2 g of benzocaine in 20 ml of 90% ethanol

Take 4 ml of the above solution and add it to 400 ml of tap water for terminal surgeries. Take 1 ml of the above and add it to 500 ml of tap water for survival surgeries.

TESTES STORAGE SOLUTION

A solution should be made containing 80% calf serum, 20% MBS and 20 mM NaCl (Sive *et al.*, 1997).

DTT SOLUTION FOR DEJELLYING

A 6.5% DTT stock solution (6.5 g DTT/100 ml water) should be made and kept at 4°C. A HEPES stock solution of 0.5 M HEPES, pH 8.9 should be made and also stored at 4°C. To make 100 ml of solution: 2 ml of DTT stock solution, 10 ml HEPES stock solution, diluted in distilled water to 100 ml to give final concentrations of 5 mM DTT and 50 mM HEPES (Sive *et al.*, 1997).

PBS 10 ×

137 mM NaCl
2.7 mM KCl
10 mM Na_2HPO_4
2 mM KH_2PO_4

The pH should be adjusted to 7.5 with the addition of HCl and/or NaOH. The solution should be autoclaved prior to use.

Protocol 7.2: X-gal staining protocol (Sive *et al.*, 1997)

1. Embryos previously injected with β-gal RNA should be monitored until they reach the desired stage of development

2. The embryos should then be washed repeatedly in 1 × PBS

3. Fixation of embryos for 1 h on ice using the following solution, for 50 ml of fixative:
   2.5 ml of 2% formaldehyde
   0.4 ml of 0.2% glutaraldehyde
   0.2 ml of 0.02% NP-40
   0.25 ml of 0.01% Na$^+$ deoxycholate
   46.45 ml of 1 × PBS

4. After 1 h in fix, embryos need to be washed several times in 1 × PBS

5. Staining with X-gal should next be conducted at approximately 30°C using the following solution:
   2.5 ml of 5 mM $K_3Fe(CN)_6$
   2.5 ml of 5 mM $K_4Fe(CN)_6$
   1.25 ml of X-gal (1 mg/ml)
   0.1 ml of 2 mM $MgCl_2$
   43.65 ml of 1 × PBS
   The length of time needed to stain the embryos varies with the depth of the β-galactosidase activity; the deeper the activity of the RNA the longer the time needed to acquire suitable staining (Sive *et al.*, 1997)

6. Embryos should be again rinsed with 1 × PBS after staining is completed. Fixation can be repeated again to allow for more stable expression of the stain (Sive *et al.*, 1997)

7. Afterwards, drain the embryos of the PBS solution and leave them stored in 100% methanol

Protocol 7.3: Overall protocol for siRNA experiment (example)

1. Buy commercially or create double-stranded siRNA that encodes for the gene of interest

2. Induce ovulation in two female *X. laevis* frogs

3. Collect the dispelled eggs

4. Isolate the testes of a male frog

5. Conduct *in vitro* fertilization

6. Dejelly the embryos and remove their vitelline membranes

7. Microinject double-stranded siRNA as well as a lineage tracer if desired

8. Screen phenotypes, or do further analysis such as RT-PCR to confirm that the level of mRNA for the gene of interest is diminished. Western blots can also be performed to confirm that protein levels from the gene of interest are decreased by the use of the siRNA

Generation of transgenic and knockdown mice with lentiviral vectors and RNAi techniques

8

Jenni Huusko, Petri I. Mäkinen, Leena Alhonen and Seppo Ylä-Herttuala

8.1 Introduction

Transgenic and knockout/knockdown mice have been invaluable for basic biological studies. Lentiviral vectors and RNAi techniques have provided new ways of producing gene-modified animals for biological studies. In this chapter, we go through the basics of generating transgenic and knockdown mice with these new techniques and give complete protocols to perform lentiviral transgenesis via two different methods.

8.2 Production of transgenic and knockdown mice

There are two basic ways to generate genetically modified mice for *in vivo* studies of gain of gene function or loss of gene function. Foreign genetic material can be introduced into embryonic stem (ES) cells grown in culture or into embryos of different stages depending on the method of gene transfer. Here, we focus on the production of transgenic and knockdown (gene-silenced) mice but not on the production of gene-disrupted (knockout) animals obtained by homologous recombination in ES cells or by any application based on nuclear transfer.

8.3 Use of ES cells

To obtain transgenic mice, the gene construct of interest is usually introduced by transfection, electroporation or viral transduction into ES cells (Hogan *et al.*, 1994). After gene transfer, analysis and selection of the cells for the desired genotype is necessary. A clonal population is then aggregated with denuded morula-stage embryos or injected into blastocyst-stage embryos of an ES cell-compatible mouse strain to allow the integration of the cells into the inner cell mass, giving rise to the tissues of the forthcoming embryo. The use of a non-clonal population of ES cells would result in chimeras with mosaicism in terms of transgene integration site and copy number. Blastocyst-stage embryos containing the manipulated ES cells are transferred to recipient females for further development. The resulting animal will always be a chimera, which is bred further to obtain pure hemizygous and eventually homozygous transgenic mice.

Production of gene-silenced mice with the aid of ES cells follows essentially the same outline regarding the processing of ES cells, embryos and chimeras. siRNAs or viral shRNA vectors inducing the generation of siRNAs can be introduced into ES cells by transfection or transduction, respectively. However, only the use of plasmid or viral vectors allowing continuous generation of siRNAs will allow the long-term silencing of the target gene desired for the ultimate *in vivo* studies (Rubinson *et al.*, 2003; Tang *et al.*, 2004). The advantage in the use of ES cells is the possibility to analyze the effect of gene manipulation in cultured cells and to select populations with different degrees of silencing before the production of chimeras and knockdown mouse lines. The obvious disadvantage is that this application is only applicable to species for which ES cells are available. Among laboratory animals, this approach is currently restricted to mice. The technique requires rather careful cell culture conditions to keep the ES cells pluripotent, and the maintenance of appropriate mouse lines for the production of compatible recipient embryos for the manipulated ES cells. Also, a lengthy backcrossing period to any desired mouse strain is needed to get rid of the chimerism. However, apart from applications in basic science using ES cell cultures or laboratory animals derived from them, genetic manipulation of ES cells may have great potential in human ES cell-based therapies.

8.4 Use of embryos

The use of ES cells in transgenesis can be bypassed with the use of preimplantation embryos. Conventional transgenic animals are routinely produced by pronuclear microinjection of 1-cell embryos (zygotes) with the gene construct of interest (Gordon *et al.*, 1980). After being transferred into the oviducts of recipient females, pups are born in due course. They are analyzed for transgenecity usually at the time of weaning. Each founder is unique as regards the gene copy number, the site of integration, and the level of transgene expression. Instead of pronuclear injection, lentiviral vectors have successfully been used to generate transgenic mice with resulting stable transgene expression (Lois *et al.*, 2002; Pfeifer *et al.*, 2002).

For short-term gene silencing experiments in preimplantation embryos, dsRNA can be injected into the cytoplasm of zygotes (Haraguchi *et al.*, 2004). In the production of gene-silenced animals, zygotes are injected with viral constructs inducing generation of siRNAs (Tiscornia *et al.*, 2003). Contrary to the above-mentioned pronuclear or cytoplasmic injection, the viral preparation is injected into the perivitelline space under the zona pellucida, not into the embryo proper, and the embryos are transferred into the oviducts of the recipient females. Another possibility is to incubate multicellular zona-free embryos in medium containing the viral construct and transfer the embryos into the oviducts or uteri of the recipient females. The perivitelline injection is technically more demanding and expensive, as it requires a microinjection apparatus. It is, however, gentle for the embryos and the visibility of intrazygotic organelles, such as pronuclei, is not critical. Moreover, it is likely to produce animals with no mosaics, as the integration of the silencing construct occurs already at the one-cell stage. Transduction of denuded two to four cell embryos by incubating them with the viral construct does not require similar technical facilities, but the removal of zona with acidic tyrode is highly toxic to embryos and handling of denuded embryos may be difficult due to their stickiness. The greatest drawback in the latter technique is the fact that resulting animals will most likely be mosaics leading to the necessity of further breeding of the founders to get pure knockdown animals. With the viral preparation, a higher titer is needed for the perivitelline injection than for the transduction of embryos in culture media. When embryos, instead of ES cells, are used as the starting material, one has to wait for the founder animals or its F_1 progeny to be born until it is possible to analyze the transgene or shRNA expression and the degree of gene silencing. Among the independent lines obtained, one can then select the ones that best serve the interests of the research.

The use of embryos in viral transgenesis, especially in the production of knockdown animals, is also applicable to other species than mouse and to any mouse strain for which embryos are relatively easily available but not suitable for pronuclear microinjection. The efficacy of viral transgenesis in general is by far much higher than the efficacy of transgenesis via pronuclear injection. The current availability and further development of vectors for tissue-specific and conditional gene expression, and silencing makes this technique an even more versatile tool in the research of living animals.

8.5 Lentivirus vectors

Lentivirus vectors (LVs) based on HIV-1 have become widely used gene transfer vehicles during the last decade. When pseudotyped with the vesicular stomatitis virus G protein, they have been shown to transduce both dividing and non-dividing cells *in vitro* and *ex vivo*, including stem cells (Miyoshi *et al.*, 1999; Naldini *et al.*, 1996). Lentiviruses can also transduce several tissues *in vivo*, for example the nervous system (Kordower *et al.*, 2000; Naldini *et al.*, 1996), hematopoietic cells (Miyoshi *et al.*, 1999; Pawliuk *et al.*, 2001), liver (Kankkonen *et al.*, 2004) and cardiomyocytes (Fleury *et al.*, 2003). LVs integrate into the genome of target cells, and gene expression from SIN-LVs is typically not silenced *in vivo* during development, which enables the use of LVs for the generation of transgenic animals (Lois *et al.*, 2002; Pfeifer *et al.*, 2002).

The LVs used for transgenesis are typically self-inactivated (SIN) third-generation vectors. In SIN-vectors, part of the viral 3'-LTR (long terminal repeat) has been deleted, preventing viral replication (Miyoshi *et al.*, 1998; Zufferey *et al.*, 1998). Additionally, third-generation LVs contain central polypurine tract (cPPT) and woodchuck hepatitis virus post-transcriptional regulatory element (WPRE), which enhance viral titers and the level of transgene expression, respectively (Follenzi *et al.*, 2000; Zufferey *et al.*, 1999). The choice of the promoter varies according to the individual needs and requirements, that is, the strength, tissue specificity, and controllability. For example, human phosphoglycerate kinase promoter, human ubiquitin-C promoter, and chicken B-actin promoter have been used in the generation of transgenic mice. The typical SIN-LV is presented in the *Figure 8.1*.

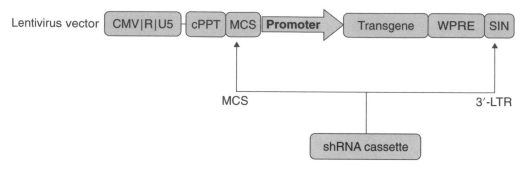

Figure 8.1

Schematic picture of the LVs used for transgenesis and two possible positions for shRNA cassette insertions.

8.6 Design of LVs for the generation of knockdown mice

8.6.1 Constitutive pol III promoters

The vectors for RNAi have traditionally contained pol III promoters, usually U6 and H1, for the expression of shRNA. The advantage of U6 and H1 promoters is that they drive very strong expression of short RNAs and are ubiquitously active in all cell types. Also, the short length and clearly defined structure of these promoters allows easy use in various expression cassettes. They have been shown to mediate long-term silencing in mouse tissues, for example, brain (Mäkinen *et al.*, 2006; Raoul *et al.*, 2005) and in *ex vivo* transduced bone marrow cells (Bot *et al.*, 2005).

The first LV-mediated generation of knockdown mouse was described in 2003, when silencing of GFP expression in transgenic mice using shRNA under H1 promoter was reported (Tiscornia *et al.*, 2003). Silencing of an endogenous gene, CD8, was also described using U6 promoter (Rubinson *et al.*, 2003). H1 promoter has been shown to generate a knockdown phenotype, even in single-copy knockdown animals (Lu *et al.*, 2004). In addition, the above-mentioned papers showed that the shRNA cassette can be inserted either upstream of the marker gene into the multiple cloning site (MCS) or into the 3'-LTR of the LV (*Figure 8.1*). In the latter case, shRNA cassette is copied into the 5'-LTR during the reverse transcription step. The duplication of the shRNA cassette might lead to an enhancement of the silencing, but on the other hand, it might disturb the expression of the marker gene. In the first case, only one copy of the shRNA cassette is present per viral integration.

Although these initial reports were promising, our own experiments and some other studies have shown that generation of knockdown mice using pol III promoters has been challenging (Carmell *et al.*, 2003; Xia *et al.*, 2006). The problem is likely linked to the strong and ubiquitous expression of shRNA. Strong overexpression leads to the saturation of the shRNA/miRNA pathway where the most likely limiting factor seems to be exportin-5, which is involved in the transfer of small RNAs from the nucleus into the cytoplasm (Grimm *et al.*, 2006; Yi *et al.*, 2005). Because of this saturation, the functions of endogenous miRNAs are disturbed, and inadequate control of certain genes can lead to toxicity, as shown by Grimm *et al.*, (2006). Also, potential induction of interferon response (Bridge *et al.*, 2003) and non-specific silencing may affect the generation of the transgenic mice.

8.6.2 Regulatable pol III promoters

To overcome the problem of saturation, conditional expression systems for pol III promoters have been developed. Ventura and colleagues (2004) used a LV-system where a modified loxP site that contained a TATA box and transcription start site for the U6 promoter replaces the 3'-end of the U6 promoter leading to its activation. When crossed with Cre-expressing mice, tissue specific conditional silencing of CD8 was detected in mouse spleen cells. A similar strategy using plasmid vector was used to knock down the bfl1/A1-gene in a conditional manner in mouse thymocytes (Oberdoerffer *et al.*, 2005).

Another strategy utilizing the Cre-loxP system with plasmid vectors was introduced by Coumoul *et al.* (2005). In this system, *neo* between two loxP sites was inserted into U6 promoter therefore inactivating it. When crossed to Cre-expressing mouse, activated U6 promoter lead to over 95% decrease in Fgfr2 expression, and consequently to embryonic lethality when Cre was expressed in germ line, or abnormal digit formation when Cre was expressed in distal mesenchyme (Coumoul *et al.*, 2005). These abnormalities were consistent with a compromised FGFR2 function.

Drug-inducible lenti-/retroviral RNAi-pol III systems have also been developed (Gupta *et al.*, 2004; Szulc *et al.*, 2006; Wiznerowich and Trono, 2003), but their application to the generation of knockdown animals has not yet been described. However, recently a drug controllable expression of shRNA under pol III promoter in mouse cells *in vivo* has been described (Szulc *et al.*, 2006). Using tetO sequence upstream to H1shRNA regulated by tTRKRAB, Tp53 gene expression was turned off and on by doxycycline in nude mice, in which LV-transduced cells were transplanted (Szulc *et al.*, 2006). Additionally, they showed that GFP expression can be controlled in mouse using the Tet-On and Tet-Off system (Szulc *et al.*, 2006).

8.6.3 Pol II promoters

The latest generation RNAi-vectors mimic the natural structure of endogenous miRNAs. The target sequence of the natural miRNA has been replaced with an appropriate target sequence, and the shRNA is expressed as a long pri-miRNA, that contains the 5'- and 3'-ends of the host miRNA (Zeng *et al.*, 2005). The benefit gained from utilizing miRNA structure is the possibility to use more flexible pol II promoters, which naturally drive the expression of miRNAs (Lee *et al.*, 2004). Compared with pol III, pol II promoters are more easily controllable, and various tissue-specific pol II promoters are also available.

Several pol II-based plasmid/viral siRNA expression systems have been described and tested *in vitro* (Silva *et al.*, 2005; Stegmeier *et al.*, 2005; Xia *et al.*, 2006; Zeng *et al.*, 2005), but so far only two reports show the generation of shRNA mice using this system (Rao *et al.*, 2006; Xia *et al.*, 2006). ShRNA against Sod2 under ubiquitin C promoter inserted in human mir-30 structure led to a sustained shRNA expression and phenotypes consistent with the Sod2 knockout mouse (Xia *et al.*, 2006). Moreover, tissue-specific knockdown of Wilms' tumor 1 transcription factor has been described using proximal promoter from mouse *Pem* gene (Rao *et al.*, 2006).

References

Bot, I., Guo, J., Van Eck, M., Van Santbrink, P.J., Groot, P.H., Hildebrand, R.B., Seppen, J., Van Berkel, T.J. and Biessen, E.A. (2005) Lentiviral shRNA silencing of murine bone marrow cell CCR2 leads to persistent knockdown of CCR2 function *in vivo*. *Blood* **106**: 1147–1153.

Bridge, A.J., Pebernard, S., Ducraux, A., Nicoulaz, A.L. and Iggo, R. (2003). Induction of an interferon response by RNAi vectors in mammalian cells. *Nat Genet* **34**: 263–264.

Carmell, M.A., Zhang, L., Conklin, D.S., Hannon, G.J. and Rosenquist, T.A. (2003) Germline transmission of RNAi in mice. *Nat Struct Biol* **10**: 91–92.

Coumoul, X., Shukla, V., Li, C., Wang, R.H. and Deng, C.X. (2005) Conditional knockdown of Fgfr2 in mice using Cre-LoxP induced RNA interference. *Nucleic Acids Res* **33**: e102.

Fleury, S., Simeoni, E., Zuppinger, C., Deglon, N., von Segesser, L.K., Kappenberger, L. and Vassalli, G. (2003) Multiply attenuated, self-inactivating lentiviral vectors efficiently deliver and express genes for extended periods of time in adult rat cardiomyocytes *in vivo*. *Circulation* **107**: 2375–2382.

Follenzi, A., Ailles, L.E., Bakovic, S., Geuna, M. and Naldini, L. (2000) Gene transfer by lentiviral vectors is limited by nuclear translocation and rescued by HIV-1 pol sequences. *Nat Genet* **25**: 217–222.

Gordon, J.W., Scangos, G.A., Plotkin, D.J., Barbosa, J.A. and Ruddle, F.H. (1980) Genetic transformation of mouse embryos by microinjection of purified DNA. *Proc Natl Acad Sci USA* **77**: 7380–7384.

Grimm, D., Streetz, K.L., Jopling, C.L., Storm, T.A., Pandey, K., Davis, C.R., Marion, P., Salazar, F. and Kay, M.A. (2006) Fatality in mice due to oversaturation of cellular microRNA/short hairpin RNA pathways. *Nature* **445**: 537–541.

Gupta, S., Schoer, R.A., Egan, J.E., Hannon, G.J. and Mittal, V. (2004) Inducible, reversible, and stable RNA interference in mammalian cells. *Proc Natl Acad Sci USA* **101**: 1927–1932.

Haraguchi, S., Saga, Y., Naito, K., Inoue, H. and Seto, A. (2004) Specific gene silencing in the pre-implantation stage mouse embryo by an siRNA expression vector system. *Mol Reprod Dev* **68**: 17–24.

Hogan, B., Beddington, R., Costantini, F. and Lacy, E. (eds) (1994) *Manipulating the Mouse Embryo: A Laboratory Manual*. Cold Spring Harbor Laboratory Press, Plainview, New York.

Kankkonen, H.M., Vähäkangas, E., Marr, R.A., Pakkanen, T., Laurema, A., Leppänen, P., Jalkanen, J., Verma, I.M. and Ylä-Herttuala, S. (2004) Long-term lowering of plasma cholesterol levels in LDL-receptor-deficient WHHL rabbits by gene therapy. *Mol Ther* **9**: 548–556.

Kordower, J.H., Emborg, M.E., Bloch, J., *et al.* (2000) Neurodegeneration prevented by lentiviral vector delivery of GDNF in primate models of Parkinson's disease. *Science* **290**: 767–773.

Lee, Y., Kim, M., Han, J., Yeom, K.H., Lee, S., Baek, S.H. and Kim, V.N. (2004) MicroRNA genes are transcribed by RNA polymerase II. *EMBO J* **23**: 4051–4060.

Lois, C., Hong, E.J., Pease, S., Brown, E.J. and Baltimore, D. (2002) Germline transmission and tissue-specific expression of transgenes delivered by lentiviral vectors. *Science* **295**: 868–872.

Lu, W., Yamamoto, V., Ortega, B. and Baltimore, D. (2004) Mammalian Ryk is a Wnt coreceptor required for stimulation of neurite outgrowth. *Cell* **119**: 97–108.

Mäkinen, P.I., Koponen, J.K., Kärkkäinen, A-M., Malm, T.M., Pulkkinen, K.H., Koistinaho, J., Turunen, M.P. and Ylä-Herttuala, S. (2006) Stable RNA interference: comparison of U6 and H1 promoters in endothelial cells and in mouse brain. *J Gene Med* **8**: 433–441.

Miyoshi, H., Blomer, U., Takahashi, M., Gage, F.H. and Verma, I.M. (1998) Development of a self-inactivating lentivirus vector. *J Virol* **72**: 8150–8157.

Miyoshi, H., Smith, K.A., Mosier, D.E., Verma, I.M. and Torbett, B.E. (1999) Transduction of human CD34+ cells that mediate long-term engraftment of NOD/SCID mice by HIV vectors. *Science* **283**: 682–686.

Naldini, L., Blomer, U., Gallay, P., Ory, D., Mulligan, R., Gage, F.H., Verma, I.M. and Trono, D. (1996) *In vivo* gene delivery and stable transduction of nondividing cells by a lentiviral vector. *Science* **272**: 263–267.

Oberdoerffer, P., Kanellopoulou, C., Heissmeyer, V., Paeper, C., Borowski, C., Aifantis, I., Rao, A. and Rajewsky, K. (2005) Efficiency of RNA interference in the mouse hematopoietic system varies between cell types and developmental stages. *Mol Cell Biol* **25**: 3896–3905.

Pawliuk, R., Westerman, K.A., Fabry, M.E., *et al.* (2001) Correction of sickle cell disease in transgenic mouse models by gene therapy. *Science* **294**: 2368–2371.

Pfeifer, A., Ikawa, M., Dayn, Y. and Verma, I.M. (2002) Transgenesis by lentiviral vectors: lack of gene silencing in mammalian embryonic stem cells and preimplantation embryos. *Proc Natl Acad Sci USA* **99**: 2140–2145.

Rao, M.K., Pham, J., Imam, J.S., MacLean, J.A., Murali, D., Furuta, Y., Sinha-Hikim, A.P. and Wilkinson, M.F. (2006) Tissue-specific RNAi reveals that WT1 expression in nurse cells controls germ cell survival and spermatogenesis. *Genes Dev* **20**: 147–152.

Raoul, C., Abbas-Terki, T., Bensadoun, J.C., Guillot, S., Haase, G., Szulc, J., Henderson, C.E. and Aebischer, P. (2005) Lentiviral-mediated silencing of SOD1 through RNA interference retards disease onset and progression in a mouse model of ALS. *Nat Med* **11**: 423–428.

Rubinson, D.A., Dillon, C.P., Kwiatkowski, A.V., *et al.* (2003) A lentivirus-based system to functionally silence genes in primary mammalian cells, stem cells and transgenic mice by RNA interference. *Nat Genet* **233**: 401–406.

Silva, J.M., Li, M.Z., Chang, K., *et al.* (2005) Second-generation shRNA libraries covering the mouse and human genomes. *Nat Genet* **37**: 1281–1288.

Singer, O., Tiscornia, G., Ikawa, M. and Verma, I.M. (2006) Rapid generation of knockdown transgenic mice by silencing lentiviral vectors. *Nat Protocols* **1**: 286–292.

Stegmeier, F., Hu, G., Rickles, R.J., Hannon, G.J. and Elledge, S.J. (2005) A lentiviral microRNA-based system for single-copy polymerase II-regulated RNA interference in mammalian cells. *Proc Natl Acad Sci USA* **102**: 13212–13217.

Szulc, J., Wiznerowicz, M., Sauvain, M.O., Trono, D. and Aebischer, P. (2006) A versatile tool for conditional gene expression and knockdown. *Nat Methods* **3**: 109–116.

Tang, F.C., Meng, G.L., Yang, H.B., Li, C.J., Shi, Y., Ding, M.X., Shang, K.G., Zhang, B. and Xue, Y.F. (2004) Stable suppression of gene expression in murine embryonic stem cells by RNAi directed from DNA vector-based short hairpin RNA. *Stem Cells* **22**: 93–99.

Tiscornia, G., Singer, O., Ikawa, M. and Verma, I.M. (2003) A general method for gene knockdown in mice by using lentiviral vectors expressing small interfering RNA. *Proc Natl Acad Sci USA* **100**: 1844–1848.

Tiscornia, G., Singer, O. and Verma, I.M. (2006) Production and purification of lentiviral vectors. *Nat Protocols* **1**: 241–245.

Ventura, A., Meissner, A., Dillon, C.P., McManus, M., Sharp, P.A., Van Parijs, L., Jaenisch, R. and Jacks, T. (2004) Cre-lox-regulated conditional RNA interference from transgenes. *Proc Natl Acad Sci USA* **101**: 10380–10385.

Wiznerowicz, M. and Trono, D. (2003) Conditional suppression of cellular genes: lentivirus vector-mediated drug-inducible RNA interference. *J Virol* **77**: 8957–8961.

Xia, X.G., Zhou, H., Samper, E., Melov, S. and Xu, Z. (2006) Pol II-expressed shRNA knocks down Sod2 gene expression and causes phenotypes of the gene knock-out in mice. *PLoS Genet* **2**: e10.

Yi, R., Doehle, B.P., Qin, Y., Macara, I.G. and Cullen, B.R. (2005) Overexpression of exportin 5 enhances RNA interference mediated by short hairpin RNAs and microRNAs. *RNA* **11**: 220–226.

Zeng, Y., Cai, X. and Cullen, B.R. (2005) Use of RNA polymerase II to transcribe artificial microRNAs. *Methods Enzymol* **392**: 371–380.

Zufferey, R., Dull, T., Mandel, R.J., Bukovsky, A., Quiroz, D., Naldini, L. and Trono, D. (1998) Self-inactivating lentivirus vector for safe and efficient *in vivo* gene delivery. *J Virol* **72**: 9873–9880.

Zufferey, R., Donello, J.E., Trono, D. and Hope, T.J. (1999) Woodchuck hepatitis virus post-transcriptional regulatory element enhances expression of transgenes delivered by retroviral vectors. *J Virol* **73**: 2886–2892.

Protocol 8.1: Mice, reagents and equipment

MICE

Four different sets of mice are needed to perform transgenesis.

- Donor females: Female mice weighing 9–12 g (aged from three to five weeks). Up to 50 embryos can be obtained from one superovulated donor mouse. We usually superovulated five to seven donors for one transgenesis session. Superovulation procedure is described below

- Breeder males: Adult males of the same strain as the donor females. At the maximum, two donor mice were mated with one breeder male, preferably one female per one male

- Recipient females: Females over 20 g of desired strain. The strain can be chosen freely and thus one can benefit from e.g. a strain with good mothering qualities. For five donors we mated at least 15 recipient females with vasectomized males

- Vasectomized males: Vasectomized males were used to induce pseudopregnancy in the recipient females. We preferred to use one recipient female per vasectomized male, two at the maximum. The vasectomy procedure is described below

Lentiviral transgenesis technique can be adjusted to any strain of mice as long as the strain has a response to the superovulation procedure. We have used CD_2F_1 hybrid mice due to their good response to the superovulation and have also started to use C57B1/6j inbred mice.

Both sets of males can be used for several months. One can have for example 2 × 15 recipient females that are bred in turn. All recipient females are not in the estrus phase of their cycle during the breeding (since there has not been any hormonal synchronization) and therefore will not mate. From 15 recipients we usually obtained six to eight females that had mated and therefore were ready to receive embryos. If the female has not mated, it will not be pseudopregnant and therefore not ready to receive embryos and carry on the pregnancy. The females that have not mated or are plugged but have not been used in embryo transfer can be used again for the next transgenesis matings right away or after a 2-week resting period, respectively.

REAGENTS

- Lentiviral construct. The titer of the virus should be at least 5×10^8 virus particles per ml (Singer et al., 2006). The production and purification of lentiviral vectors is described in detail elsewhere (Tiscornia et al., 2006)

- Pregnant mare's serum gonadotropin (PMSG) (Sigma G4877). 2000 IU PMSG is used to mimic follicle-stimulating hormone (FSH), which increases the number of developing follicles

- hCG (Pregnyl®). 5000 IU hCG mimics luteinizing hormone (LH), which induces ovulation, i.e. the rupture of mature follicles

- M2 Medium (Sigma M7167)

- M16 Medium (Sigma M7792 + 100 × penicillin streptomycin)

- Mineral oil (Sigma M8410)

- Hyaluronidase solution in M2 Medium (final concentration 0.001 g/ml, Sigma H3506)

- Acidic Tyrode Solution for removal of the zona pellucida: 0.800 g NaCl, 0.020 g KCl, 0.024 g $CaCl_2 \cdot H_2O$, 0.010 g $MgCl_2 \cdot H_2O$, 0.100 g glucose and polyvinylpyrrolidone (PVP) in 100 ml water. Adjust pH to 2.5, store at $-20°C$ (Hogan *et al.*, 1994)

- Silicone solution for siliconizing micro slides for injection chambers (Sigma SL2)

- Anesthetic (Isoflurane: induction 4–5% , maintenance 2–3%)

- Analgetic (Carprofen: 5 mg/kg subcutaneous)

REAGENTS FOR ANALYZING THE PUPS

- Digestion buffer: 0.5 ml 1 M Tris (pH 8), 0.1 ml 0.5 M EDTA, 1 ml 10% SDS, 0.5 ml 5 M NaCl, 0.1 ml Proteinase K (10 mg/ml), 7.8 ml H_2O. 10 ml is enough for 25 samples. Digestion buffer should be prepared fresh

- DNA polymerase (Finnzymes F-505L)

- Deoxyribonucleotide triphosphate (dNTP) mix (Finnzymes F-560L)

- Primers for amplification (depending on the transgene/siRNA)

EQUIPMENT

- Microscopes for embryo collection and embryo transfer (Olympus SZH-ILLK, Olympus Optical Co., Ltd.)

- Incubator with humidified 5% CO_2 atmosphere

- Micromanipulator = 'joystick' (Eppendorf Micromanipulator 5171)

- Micrometer syringe (Narishige)

- Microinjector (Eppendorf Microinjector 5242)

- Inverted microscope (Zeiss Axiovert 35M)

- Micropipette puller for injection needles (Sutter Instrument Co., Model P-97)

- Glass slides and small cover slides for injection chambers

- Glass Pasteur pipettes (Brand GMBH & Co.)

- 3.5-cm culture dishes (Sarstedt)

- Glass capillaries to produce the injection needles (length 10 cm, diameter 1 mm) and holder capillaries (length 15 cm, diameter 1 mm) (World Precision Instruments)

- Bunsen burner

- Instruments: small scissors (14088-10), 2 × fine-point sharp forceps for harvesting (11241-30), 2 × blunt forceps (11150-10), Tissue clamp (18050-12), 2 × fine-point sharp forceps for transfer (11295-10), needle holders (13008-12) (Fine Science Tools)

- 5-0 Mersilk Suture for embryo transfer (absorbable) (Johnson-Johnson Intl)

- 5-0 Suture for vasectomy procedure (Johnson-Johnson Intl)

Protocol 8.2: Setting up capillaries, injection needles, injection chambers and preparations for transgenesis

EMBRYO-HARVESTING CAPILLARIES

Make embryo-harvesting capillaries according to Hogan *et al.* (1994). Cut the tip to approximately 4 cm in length, examine with microscope that the tip has broken evenly, and keep it in a place where the tip does not break. These pipettes can be prepared in large numbers in advance.

EMBRYO-TRANSFER CAPILLARIES

Prepare embryo-transfer capillaries according to Hogan *et al.* (1994). Cut the tip to approximately 4 cm in length, examine with microscope that the tip has broken evenly, and keep it in a place where the tip doesn't break. These pipettes can be prepared in large numbers in advance.

INJECTION CHAMBERS

Embryos are kept in an injection chamber during microinjection. To make an injection chamber, siliconize microslides by pipetting a drop of silicone solution in the middle of the microslide swept with ethanol. Let the silicone solution dry and store the slides in a refrigerator. Siliconized microslides can be prepared in large numbers in advance. When starting the procedure, sweep one siliconized microslide with ethanol. Set an approximately 35 µl drop of M2 medium on the microslide. Sweep a smaller cover glass with ethanol and place stripes of white vaseline and tooth wax mixture with a syringe on two sides of the cover glass. Place pieces of Pasteur pipette tips cut to the length of the cover glass on top of the vaseline-wax stripes. Then place another stripe of vaseline-wax mixture on top of the Pasteur pipette tips. Place the cover glass on top of the M2 medium drop on the siliconized microslide and push down so that the medium drop touches the cover glass. This way a chamber is formed so that the two glasses are approximately 1 mm apart and the drop of M2 medium is between the glasses. Tape this system to two microslides that serve as props. Finally, fill the chamber between the siliconized microslide and the cover glass with mineral oil. (*Figure 8.2.*)

INJECTION NEEDLES

Use glass capillaries of 1 mm in diameter and 10 cm in length to make injection needles. Pull the needles with a Sutter micropipette puller by using the following cycle settings: heat 445, pull 200, velocity 60, and time 150. The injection needles should not be prepared a long time beforehand because the moisture condensing in them might become a problem. Prepare 8–10 injection needles for one microinjection session to be able to change the needle in case of clogging or breakage.

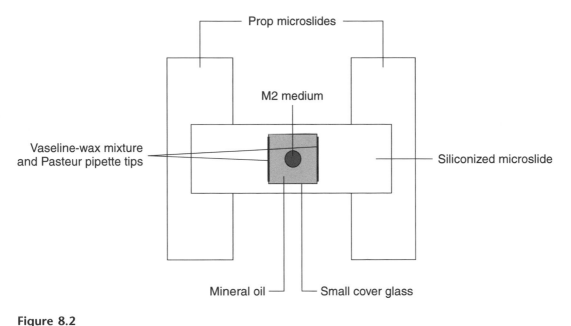

Figure 8.2

Injection chamber.

EMBRYO HOLDER CAPILLARIES

Holder capillaries are used to hold an embryo still while it is microinjected. Use glass capillaries of 1 mm in diameter and 15 cm in length to make the holder capillaries. Keep a capillary in a Bunsen burner flame until it softens from the middle of its length. Withdraw from the flame and pull apart from both ends to make the outer diameter to 100–120 μm. Cut the pulled capillary in half and melt the cut end near the Bunsen burner flame to round the sides and make the final hole in the tip to approximately 15 μm in diameter. Use a microscope to examine the holder tip before use.

VASECTOMY

Vasectomized males are used to cause pseudopregnancy in the recipient females, as these males do not deliver any sperm in their semen. Vasectomized males can be produced by either cutting or cauterizing the vas deferens such that its reconnection is unlikely. We routinely use the cutting procedure.

The vas deferens can be exposed either via a midline ventral incision in the abdominal wall or an approximately 5 mm incision into each scrotal sac. The scrotal sac method is the preferred one and should be used instead of the laparotomy.

Vasectomy is done to young males that have reached sexual maturity (10–11 weeks of age). Anesthetize animals by using a suitable anesthetic (http://www.uku.fi/vkek/ohjeistusta/Nukutus-kipu-postoper/). Make one, approximately 1-cm midline incision in the skin of the scrotum. Blunt dissect a small area subcutaneously to either the left or right of the incision such that an outline of a testis can be seen through the body wall. Then make an approximately 5-mm incision in the body wall to one side in line with the base of the testis. Locate the cauda epididymis and the pale vas deferens and dissect away a prominent blood vessel leading from it so that a loop is formed. Ligate the vas deferens such that a 5–7-mm section can be cut between the ligations. Make a single stitch to the body wall and repeat the procedure for the other side. Finally, stitch the skin. Give post-operative analgesia according to the institutional guidelines (http://www.uku.fi/vkek/ohjeistusta/Nukutus-kipu-postoper/).

Following surgery, allow the males to recover for a period of three weeks. This also ensures that no residual sperm remains in the proximal part of the vas deferens. After the recovery period, test mate the vasectomized males to confirm the success of the vasectomy. The vaginal plug rate should not be affected by the vasectomy but it is good practice to keep a plugging record for each vasectomized male to monitor their performance. Vasectomized males can be used weekly and are expected to plug females regularly for at least one year.

SUPEROVULATION

Sexual maturity of the donor mice is the major factor affecting the number of egg cells that are ovulated and can be collected for manipulation. Depending on the strain used, the best response to superovulative hormones is usually obtained during the prepubescent stage of development. Mice at the precise age and weight to perform superovulation may not be readily available from commercial suppliers. Therefore, one usually has to maintain a breeding colony to get females at the optimal age (Hogan et al., 1994).

Superovulate CD_2F_1 females weighting 9–12 g (aged three to five weeks of age) by using PMSG and hCG hormones to gain the maximum number of embryos. The timing of the hormone injections relative to each other and the light cycle of the mouse facility affects the uniformity and the number of egg cells gained from the donor mice (Hogan et al., 1994). In our facility, the hormones are injected between 11 a.m. and 1 p.m.

Inject 5 IU of PMSG intraperitonally in 0.1 ml volume. The day when PMSG is given is referenced as the day 1. Exactly 48 h after PMSG injection, inject 5 IU of hCG intraperitonally in 0.1 ml volume. On day 3 (the same day as the hCG injection is given), mate these females overnight with fertile breeder males to yield fertilized egg cells for manipulation. Notice that the stress induced to the donor females while the hormone injections are given can have a major impact on the number of egg cells developed and ovulated. Therefore, it is important to have an experienced animal technician who performs the superovulation procedure so that the number of animals needed can be kept at the minimum.

Protocol 8.3: Direct microinjection of the viral construct to the subzonal space (= perivitelline space) of a fertilized egg cell

1. Superovulate the donor mice as described above and mate with the breeder males. On day 3, thaw aliquots of M2 and M16 media and prepare dishes with M16 drops: make one dish with three 25-μl drops of M16 medium and two dishes with five 25-μl drops of M16 medium. Fill the dishes with mineral oil so that the drops are covered and place them into a CO_2 incubator at 37°C. Keep the thawed M16 medium in the incubator and M2 medium in a refrigerator until the next day. Check that embryo-harvesting and embryo-transfer capillaries as well as embryo-holder capillaries are ready

2. On day 4, harvest single-cell embryos in the morning (preferably 24 h after the male and female have been introduced). Sacrifice the donor females by cervical dislocation and collect the oviducts surgically by cutting from the fat pad connected to the kidneys and in the upper part of the uterus' horn. Place these oviduct–ovary bundles immediately in a dish containing prewarmed M2 medium. With the aid of a microscope, clean the oviducts free from ovaries, parts of uterus and adipose tissue. The ampulla (enlargement of the oviduct) should be visible if the ovulation has taken place. Move oviducts to another dish with M2 medium and release the embryos by breaking the ampulla with fine-point forceps. Collect the embryos with an embryo-harvesting capillary connected to mouth pipette tubing and transfer them to hyaluronidase solution, which will enzymatically digest the cumulus cells surrounding the embryos. Notice that hyaluronidase is toxic to embryos and they should not be kept in the solution any longer than necessary, i.e. no more than a few minutes. Rinse embryos through the dish with three drops of M16 medium and then divide them to the dish with five drops of M16 medium (20–60 embryos per drop). Place the dish into an incubator (+37°C, 5% CO_2) for a few hours

3. Prepare two injection chambers and pull the injection needles with the Sutter pipette puller

4. Thaw the viral construct with a titer of at least 5×10^8 virus particles per ml (Singer et al., 2006) (e.g. 50 μl in a microcentrifuge tube) and centrifuge for 5 s at full speed to pellet possible cellular

debris that might clog the tip of the injection needle and remove the supernatant to another tube. To fill the injection needles, place two or three injection needles with their tip upwards into the tube containing the viral construct. The capillary action will raise the viral construct to the tip

5. Transfer the embryos from one drop of the dish in the incubator into the injection chamber so that the uninjected embryos are in the upper part of the M2 drop in the injection chamber

6. Prepare the micromanipulator. Place an embryo-holder to the left side into a micrometer syringe, fill it with mineral oil and bring it to the injection chamber placed in the middle of the plate of an inverted microscope. Then place a filled injection needle to the right side into a microinjector and bring it also to the chamber. Use the inverted microscope to get the embryos, the holding capillary, and the injection needle visible. The injection needle has to be cut from the tip by touching the end of the holding capillary. Otherwise the viral construct will easily clog the needle

7. One by one, hold the embryos with the holding capillary (use the suction of the oil in the syringe to hold the embryo still) and inject the viral construct under the zona pellucida. The microinjector uses pressure to inject the viral construct. After piercing the zona, switch on pressure to inject viral construct to the perivitelline space. Keep the pressure on for 3–10 s, depending on how the zona seems to give in. Inject only the fertilized embryos (two polar bodies and two visible pro-nuclei). Bring the injected embryos down to the left corner of the M2 drop in the injection chamber and non-fertilized or lysed embryos down to the right side of the drop. Use smaller magnification (50×) when transferring the embryos and larger magnification (400×) while injecting the embryos. Once all embryos in the chamber have been handled, transfer them to a new dish with five drops of M16 and incubate overnight (+37°C, 5% CO_2). Then repeat the same procedure for all collected embryos

8. On day 4, mate recipient females with vasectomized males to induce pseudo-pregnancy

9. On day 5, check the embryos. They should be at the two-cell stage. Transfer the two-cell stage embryos to the oviducts of 0.5-days postcoitum (DPC) pseudopregnant recipients, pseudopregnancy is concluded from a vaginal plug, which indicates that mating has taken place. Anesthetize recipient female, give analgesic, and wipe its back with ethanol. Make a parting in the fur and make longitudinal incision of approximately 1 cm to the skin. Slide the skin to the side by blunt dissecting until the fat pad and the ovary are visible through the body wall. Make a cut through the body wall and lift the ovary and oviduct out by the fat pad connected to the ovary. Pierce the bursa between the ovary and the oviduct under the microscope and find the opening of the

Figure 8.3

PGK-GFP founder mouse (right) and a litter-mate control mouse under ultraviolet light.

oviduct. Use an embryo-transfer capillary connected to a mouth pipette tubing to transfer the embryos. Transfer 15–20 embryos per one oviduct. Place the embryos in the capillary between two air bubbles. After successfully transferring the embryos, two air bubbles should be visible in the ampulla of the recipient female. After the embryos have been transferred, gently reposition the oviduct, ovary and fat pad with blunt forceps, and sew the body wall with one stitch. Repeat the procedure on the other side and close the incision with stitches

The schedule of Protocol 8.3 is shown in *Table 8.1*.

Table 8.1 Schedule of the viral microinjection procedure

Day 1	Day 2	Day 3	Day 4	Day 5
PMSG injection (5 IU in 0.1 ml ip.)		hCG injection (5 IU in 0.1 ml ip.) Mating of donor mice with breeder males	Harvesting the embryos from the donor mice and microinjection with the viral construct. Mating of recipient females with vasectomized males	Transfer of two-cell embryos into the oviducts of pseudo-pregnant recipient females

Protocol 8.4: Zona pellucida removal and lentiviral transduction

This protocol can be used if no micromanipulator is available. The procedure is somewhat easier since there is no need to work with the micromanipulator apparatus. However, the procedure is more stressful to the embryos and one should be prepared for a larger loss of embryos than by using the direct microinjection protocol. The zona pellucida removal protocol also requires more work, since the embryos have to be handled in groups of no more than five.

1. Superovulate the donor mice as in the previous protocol

2. On day 5, thaw aliquots of M2 and M16 media, prepare one dish with three 25-µl drops of M16 medium and one dish with five 25-µl drops of M16 medium

3. Harvest two to four cell embryos on day 6, 72 h after the donor mice were introduced with the breeder males. Collect the oviducts as in the previous protocol, but now flush the oviducts with M2 medium by using a 30G needle, since two to four cell embryos will not be in the ampulla anymore. The hyaluronidase step is not required since cumulus cells are not present on the surface of the multicellular embryos. Perform the rinsing step as described.

The schedule of Protocol 8.4 is shown in *Table 8.2*.

Table 8.2 Schedule of the zona pellucida removal and lentiviral transduction

Day 1	Day 2	Day 3	Day 4	Day 5	Day 6	Day 7	Day 8
PMSG injection (5 IU in 0.1 ml ip.)		hCG injection (5 IU in 0.1 ml ip.) Mating of the donor mice with breeder males		Mating of the recipient females	Harvesting two to four cell embryos, the removal of the zona pellucida and viral co-incubation		Transfer of blastocyst stage embyos

4. Remove the zona pellucida by transferring the embryos in groups of five or less to a drop of acidic tyrode solution. Observe the embryos under a microscope during the incubation (30 s to 1 min) and remove them immediately after the first signs of zona removal. Note that acidic tyrode is very toxic to the embryos, and for optimal survival, the incubation should be kept as short as possible. After the zona has been removed, rinse the embryos through six drops of M2 and again through three drops of M16

5. For transduction, dilute the viral construct in M16 with a final volume of 2 ml and a minimum of 1×10^6 viral particles per ml (Singer *et el.*, 2006). Make 10-µl droplets under mineral oil and place one embryo per droplet. Incubate the embryos in these viral droplets for 48 h, after which they should be at the blastocyst stage. It is not necessary to refresh the droplets during the incubation

6. Transfer the blastocysts to the uteri of 2.5-DPC pseudopregnant females. Work as in the previous protocol but instead of piercing the bursa, use a 30G needle to pierce the muscular uterine wall and after removing the needle, place an embryo-transfer capillary connected to a mouth pipette tubing with 10–15 blastocysts in the capillary into the hole and transfer the embryos. Finish as in the previous protocol

ANALYSIS OF THE PUPS

After a successful embryo transfer, pups are born 19 days after the transfer to the oviducts in the direct microinjection protocol and 17 days after the transfer to the uterus in the zona pellucida removal protocol. At three to four weeks of age, the pups are weaned and identified by punching of the ears. The ear punches are used for genotyping. Lyse tissue pieces in the lysis buffer overnight at +37°C or approximately 4 h at +55°C in a shaker. Extract genomic DNA and run a PCR with primers specific for the transgene or the lentiviral backbone. Load 10-µl samples on agarose gel containing ethidium bromide and run electrophoresis.

EXPERIENCE FROM THE PROTOCOLS

Our laboratory uses mostly the direct microinjection method for the production of transgenic and knockdown mice with lentivirus. In our hands, this protocol has been far less traumatic to the embryos. We have obtained a good transgenesis rate after the microinjection procedure (*Table 8.3*). The survival rate of the embryos to two-cell stage after microinjection was remarkably high, ranging from 85% to 100%. The survival rate of embryos after the zona removal procedure and incubation with lentiviral vector (Lois *et al.*, 2002) has not been as good as with the microinjection

Table 8.3 Experience from direct microinjection of lentiviral constructs to the perivitelline space of fertilized egg cells

Construct	Number of injected embryos (Microinjection sessions)	Number of two-cell stage embryos (% of lysed embryos)	Embryo transfer procedures (number of embryos transferred)	Births (% of embryos transferred)	Transgenic pups (% of births)
PGK-GFP	98 (1)	95 (3.1)	2 (68)	6 (8.8)	5 (83.3)
Tie1-GFP	240 (2)	228 (5.0)	5 (192)	3 (1.6)	2 (66.7)
SM22 α-GFP	173 (2)	147 (15.0)	4 (132)	6 (4.5)	3 (50)
PGK-VEGF-D	145 (1)	132 (8.9)	3 (114)	4 (3.5)	4 (100)
Tie1-VEGF-D	22 (1)	22 (0)	1 (22)	2 (9.1)	2 (100)
hU6-shVEGF-D	130 (1)	124 (4.6)	3 (111)	4 (3.6)	2 (50)
hU6-shApoB100	208 (2)	190 (8.7)	5 (190)	12 (6.3)	10 (83.3)

procedure (*Table 8.4*). Many of the collected embryos did not survive from the acidic tyrode treatment, and embryos with denuded zona were often delayed in their development or did not develop to blastocyst stage at all. Although embryos could also be transferred at the morula stage, we found them not very viable after the removal of the zona and prolonged *in vitro* development. Also, it should be taken into account that founder mice (F_0) generated by the zona pellucida removal protocol are likely to be mosaic, because single-cell embryos denuded of zona will not efficiently develop to blastocysts and the transduction has to be performed with two- to four-cell embryos (Singer *et al.*, 2006). Therefore, the founder mice should be crossed with a non-transgenic mouse to achieve complete germ line transmission to the progeny. With the direct microinjection protocol, this problem usually does not occur, since embryos are injected with viral construct at the one cell stage. Founder mice will more likely then show the full phenotype caused by the transgene-induced overexpression or knockdown. Knockdown transgenesis is also a very valuable method to investigate genes that are lethal when knocked out completely. To achieve RNAi-mediated knockdown, insertion of only a single copy of the transgene is required, whereas a traditional knockout involves removing the gene from both chromosomes. Thus, generation of a knockdown mouse is much faster than production of a knockout mouse. By changing the silencing efficiency of the siRNA construct, it is possible to investigate changes in the phenotype caused by graded degrees of silencing of a particular gene. According to our experience, several knockdown mice regardless of the gene of interest have been born significantly smaller than their littermates and have also shown delayed development. This has been suggested to be due to the oversaturation of the cellular microRNA/shRNA pathways (Grimm *et al.*, 2006). Part of the problem is related to the use of strong pol III promoters, such as U6. Thus, it is advisable to have as low copy number of the siRNA as possible, preferably below three to five.

Table 8.4 The survival rate of embryos after removal of the zona pellucida and incubation with lentiviral vector

Number of embryos in zona removal	Number of embryos in transduction (% of number of embryos)	Number of morulae (% of transduced embryos)
40	40 (100)	20 (50)
174	128 (73.5)	87 (68.0)
142	114 (80.3)	37 (32.5)
80	46 (57.5)	24 (52.2)
162	110 (67.9)	48 (43.6)
198	87 (43.9)	31 (35.6)
231	104 (45.0)	63 (60.6)
218	107 (49.1)	45 (42.1)
220	109 (49.5)	79 (72.5)
255	105 (41.2)	41 (39.0)

GFP-MICE

We generated GFP-expressing mice under different promoters by using the direct microinjection protocol (*Figure 8.3*, between pages 106 and 107). These mice showed different degrees of GFP expression. The time scale of one transgenesis session is shown in *Figure 8.4* and a flowchart of the method is shown in *Figure 8.5*. All transgenic mice generated with the direct microinjection protocol have been fertile and have passed the transgene to the next generations. Thus, lentiviral transgenesis is an effective and safe method for the production of transgenic mice.

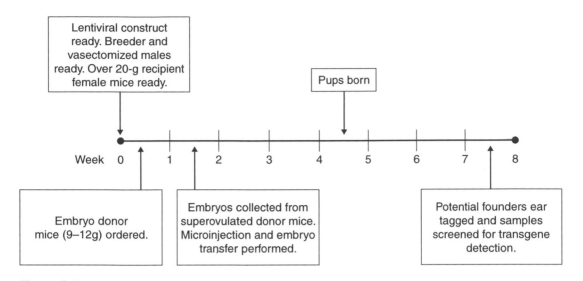

Figure 8.4

Timescale of a transgenesis session.

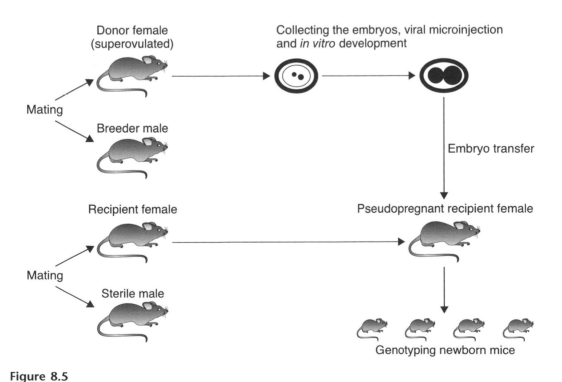

Figure 8.5

Flow-chart picture of the generation of transgenic mice by the direct microinjection protocol.

RNAi in fungi

9

Hitoshi Nakayashiki

9.1 Introduction

9.1.1 The discovery of quelling in *Neurospora*

The story of RNAi in fungi began with an interesting finding reported by Romano and Machino in 1992, whereby gene expression was shown to be interfered with by transformation with homologous sequences in the fungus *Neurospora crassa* (Romano and Machino, 1992). The gene inactivation was spontaneously reversible and involved the silencing of both transgenes and endogenous genes. This phenomenon was termed 'quelling'. An intriguing feature of quelling is that it is dominant in heterokaryons with nuclei from quelled and wild-type strains, suggesting that a mobile signal acts *in trans* to cause silencing (Cogoni *et al.*, 1996). A series of studies on quelling-deficient (*qde*) mutants of *N. crassa* (Cogoni and Macino, 1997) has provided evidence of the molecular link between quelling and RNA-mediated gene-silencing mechanisms in other organisms, which include post-transcriptional gene silencing (PTGS) or co-suppression in plants and RNAi in animals. Moreover, the *qde-1* mutant was shown to be defective in an RNA-dependent RNA polymerase (RdRP) (Cogoni and Macino, 1999), which is consistent with the findings that the SDE1/SGS2 gene in *Arabidopsis* (Dalmay *et al.*, 2000; Mourrain *et al.*, 2000) and ego-1 gene in *Caenorhabditis elegans* (Smardon *et al.*, 2000), both of which encode RdRP, are required for PTGS and RNAi, respectively. Similarly, the protein product encoded by the second qde gene, *qde-2*, was shown to be a member of the Argonaute family, which is an essential component of the PTGS and RNAi pathways (Cogoni and Macino, 2000; Fagard *et al.*, 2000). Furthermore, involvement of two dicer proteins (Dcl-1 and Dcl-2) and siRNA biogenesis in the quelling pathway was recently demonstrated (Catalanotto *et al.*, 2002, 2004). Thus, the genetic and biochemical evidence suggests that quelling belongs to the broad category of RNA-mediated gene-silencing mechanisms, exemplified by RNAi, that are evolutionary conserved in most eukaryotes.

9.1.2 Meiotic silencing by unpaired DNA (MSUD), a novel gene-silencing phenomenon in *Neurospora*

While quelling operates during the vegetative phase of growth, a novel silencing phenomenon that occurs for a limited period during the sexual phase from an early stage of meiosis after karyogamy to ascospore maturation has been identified in *N. crassa*; this is termed meiotic silencing by unpaired DNA (MSUD) (Shiu *et al.*, 2001). In contrast to quelling, which is evolutionarily conserved, operation of MSUD has, to date, been demonstrated only in *Neurospora*. MSUD was first proposed in a study on meiotic transvection of the *Asm-1* gene (Aramayo and Metzenberg, 1996), in which it was suggested that during meiosis some sensing mechanism operates to detect the presence of a pairing partner on the homologous chromosome for each allele in the zygote. MSUD silences the expression of genes that exist in one parental chromosome but not in its pairing partner, thus causing unpaired DNA during meiosis. Interestingly, MSUD affects not only the unpaired gene but also any copy of the unpaired sequence in the genome, whether the additional copies are paired or not. This suggests that a mobile trans-acting signal is involved in MSUD, as is the case in quelling.

Genetic screens for suppressors of meiotic transvection and MSUD have revealed a marked similarity between MSUD and quelling. Three suppressor loci of MSUD, sad-1 (suppressor of ascus dominance-1), sms-2 (suppressor of meiotic silencing-2) and sms-3, have been characterized and found to encode paralogs of QDE-1 (RdRP), QDE-2 (Argonaute) and DCL-2 (Dicer), respectively (Galagan *et al.*, 2003; Lee *et al.*, 2003; Shiu *et al.*, 2001). This clearly indicates that MSUD is based on a molecular mechanism very similar to that of quelling, and therefore, RNAi. An intriguing point is that different sets of protein components are used for MSUD and quelling with some redundancy between dicer-like proteins, DCL-2 and SMS-3 (DCL-1) (Galagan *et al.*, 2003). Therefore, it appears that *Neurospora* has two separate RNAi-related pathways in two different phases of its life cycle; quelling for the vegetative phase and MSUD for the sexual phase (*Figure 9.1*).

9.1.3 RNAi as a genetic tool in fungi

RNAi reportedly occurs in all four eukaryotic kingdoms: *Animalia*, *Plantae*, *Fungi*, and *Protista*, but not in the prokaryotic kingdom *Monera*. BLAST searches against the public genome databases of 15 fungal species have shown that most, but not all, fungi possess multiple paralogs of RNAi-related proteins such as RdRP, Argonaute, and Dicer in their genome (Nakayashiki *et al.*, 2006). In fact, RNAi has been experimentally demonstrated in a variety of fungal species encompassing major fungal taxa such as *Ascomycota* (Fitzgerald *et al.*, 2004; Hamada and Spanu, 1988; Hammond and Keller, 2005; Kadotani *et al.*, 2003; McDonald *et al.*, 2005; Mouyna *et al.*, 2004; Rappleye *et al.*, 2004; Spiering *et al.*, 2005), *Basidiomycota* (de Jong *et al.*, 2006; Liu *et al.*, 2002; Namekawa *et al.*, 2005), and *Zygomycota* (Nicolas *et al.*, 2003; Takeno *et al.*, 2005), as well as the fungus-like *Oomycota* (Latijnhouwers *et al.*, 2004; van West *et al.*, 1999) and *Myxomycete* (Martens *et al.*, 2002) (*Table 9.1*). Therefore, it is likely that the RNAi technology can be applied as a genetic tool in most fungal species (Nakayashiki, 2005). Interestingly, however, the information obtained from fungal genome projects also suggests that some fungi such as *Saccharomyces cerevisiae* (budding yeast), *Candida albicans*, and *Ustilago maydis* may have lost the RNAi machinery during evolution (Aravind *et al.*, 2000; Nakayashiki *et al.*, 2006), indicating that RNAi and its related pathways are not essential for growth and development in the life cycle of these fungal species.

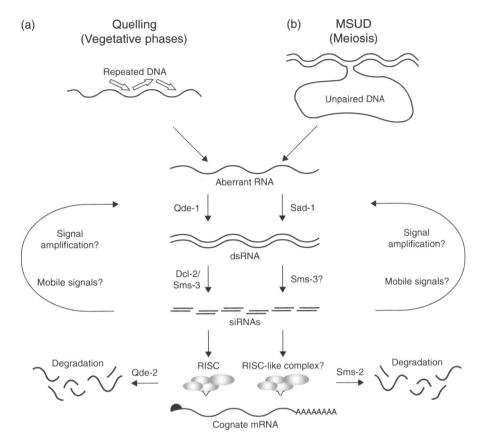

Figure 9.1

A proposed model of two RNA silencing pathways in *Neurospora crassa*. (A) During the vegetative phases of the *N. crassa* life cycle, repeated sequences in the genome can induce quelling. In this pathway, dsRNA produced by Qde-1 RNA-dependent RNA polymerase is diced into siRNAs by the action of Dcl-2 and Sms-3. The siRNAs guide degradation of cognate mRNA after their incorporation into RISC, where Qde-2 is one of the components. (B) During meiosis, a DNA fragment that has failed in pairing (unpaired DNA) triggers the second RNA silencing pathway in *N. crassa*, called MSUD. Mechanisms of silencing in MSUD are supposed to be quite similar to those in quelling except that MSUD uses a different set of silencing protein components (paralogs) from those in the quelling pathway.

For functional genomics research in fungi, the RNAi approach has potential advantages over conventional gene knockout strategies. In model fungi such as *S. cerevisiae, S. pombe, A. nidulans,* and *N. crassa*, the gene targeting method employing homologous recombination is routinely used for verifying gene function. This approach has been successfully employed even in genome-scale analysis of gene function in *S. cerevisiae* (Giaever *et al.*, 2002), where homologous recombination occurs at an extremely high efficiency. However, the efficiency of homologous recombination varies considerably among fungal species. It has been reported in *Septoria lycopersici* that the targeting efficiency of homologous recombination was less than 1% in attempts to mutate the tomatinase gene (Martin-Hernandez *et al.*, 2000). In

Table 9.1 RNAi in fungi and fungus-like organisms

Species	RNAi trigger	Transformation	Reference
Ascomycota			
Neurospora crassa	Homologous transgene	PEG-mediated method	Romano and Machino, 1992
N. crassa	IR*	PEG-mediated method	Goldoni *et al.*, 2004
Cladosporium fulvum	Homologous transgene	PEG-mediated method	Hamada and Spanu, 1988
Magnaporthe oryzae	IR	PEG-mediated method	Kadotani *et al.*, 2003
Venturia inaequalis	IR	PEG-mediated method	Fitzgerald *et al.*, 2004
Aspergillus fumigatus	IR	PEG-mediated method	Mouyna *et al.*, 2004
Histoplasma capsulatum	IR	Electroporation	Rappleye *et al.*, 2004
Aspergillus nidulans	IR	PEG-mediated method	Hammond and Keller, 2005
Fusarium graminearum	IR	PEG-mediated method	McDonald *et al.*, 2005
Neotyphodium uncinatum	IR	Electroporation	Spiering *et al.*, 2005
Basidiomycota			
Cryptococcus neoformans	IR	Electroporation	Liu *et al.*, 2002
Coprinus cinereus	IR	Lithium acetate method	Namekawa *et al.*, 2005
Schizophyllum commune	IR	PEG-mediated method	de Jong *et al.*, 2006
Zygomycota			
Mucor circinelloides	Homologous transgene	PEG-mediated method	Nicolas *et al.*, 2003
Mortierella alpina	IR	Microparticle bombardment	Takeno *et al.*, 2005
*Oomycota***			
Phytophthora infestans	Homologous transgene	PEG-mediated method***	van West *et al.*, 1999
P. infestans	Homologous transgene	Electroporation	Latijnhouwers *et al.*, 2004
P. infestans	dsRNA	Lipofectin-mediated transfection	Whisson *et al.*, 2005
Myxomycete (Slime mold)**			
Dictyostelium discoideum	IR	Electroporation	Martens *et al.*, 2002

*IR, hairpin RNA or inverted repeat RNA-expressing plasmid
** Fungus-like organisms
***Lipofectin was added to increase transformation efficiency

addition, the majority of fungi consist of multicellular or multinuclear hyphae, some of which have two or more genetically different nuclei in a common cytoplasm, referred to as a heterokaryon. These characteristics of fungi make gene targeting complicated and inefficient. Since RNAi is locus-independent and mediated by a mobile trans-acting signal in the cytoplasm, it can be applicable to any fungal species regardless of gene targeting efficiency and karyotype.

9.2. RNAi strategies in fungi

9.2.1 RNAi using a hairpin RNA-expressing plasmid

'Canonical' quelling in *N. crassa* can be induced by transformation with promoter-less transgenes (partial coding sequence) homologous to an endogenous target (Romano and Macino, 1992). However, recently it was shown that constructs expressing self-complementary hairpin RNA induce more efficient and stable silencing than canonical quelling in *N. crassa* (Goldoni *et al.*, 2004). Actually, to date, RNAi in fungi has mostly been induced by plasmid constructs that express hairpin RNA, often with an intron sequence at the loop structure. This approach has been successfully used to trigger RNAi in a variety of fungal species (*Table 9.1*).

To facilitate construction of a hairpin RNA-expressing plasmid by PCR-based cloning, the versatile vector pSilent-1 was developed for ascomycete fungi (Nakayashiki *et al.*, 2005) (*Figure 9.2A*). It carries a hygromycin resistance cassette and transcriptional unit for hairpin RNA expression with a spacer consisting of a cutinase gene intron from the rice blast fungus *Magnaporthe oryzae*. pSilent-1 is available upon request at Fungal Genetic Stock Center (http://www.fgsc.net/). Two *M. oryzae* endogenous genes, MPG1 and polyketide synthase gene, in addition to the model gene GFP were successfully silenced at various degrees by pSilent-1-based vectors in 70–90% of the resulting transformants (Nakayashiki *et al.*, 2005). It was also reported that pSilent-1 can be used to induce RNAi in another ascomycete fungus, *Colletotrichum lagenarium*, at a slightly lower efficiency than in *M. oryzae*. Therefore, pSilent-1 may serve as an efficient reverse genetic tool for exploring gene function in ascomycete fungi.

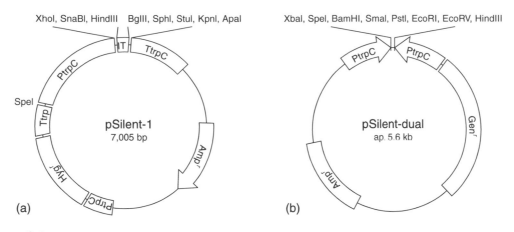

Figure 9.2

Schematic representation of the silencing vectors pSilent-1 and pSilent-dual. (A) A map of pSilent-1. Amp^r, ampicilin-resistant gene; Hyg^r, hygromycin-resistant gene; IT, intron 2 of the cutinase gene from *Magnaporthe oryzae*; PtrpC, *Aspergillus nidulans* trpC promoter; TtrpC, *A. nidulans* trpC terminator. (B) A map of pSilent-dual. Gen^r, geneticin-resistant gene; Pgpd, *A. nidulans* gpdA promoter.

9.2.2 RNAi using an opposing-dual promoter system

Recently, another type of silencing vector, which possesses multi-cloning sites between two opposing promoters, has been developed and shown to induce RNAi in fungi (*Figure 9.2B*). In this system, sense and antisense RNA of the target gene, which is expected to form dsRNA in the cell, are transcribed independently under the control of the two opposing RNA polymerase II promoters. While pSilent-1-like vectors, which require two steps of orientated cloning, are not feasible on a global scale, the opposing promoters system, which allows one-step non-oriented cloning in construction of silencing vectors, could provide a high-throughput method for RNAi in fungi. Rappleye *et al.* (2004) reported that a silencing vector of this type induced only moderate silencing of GFP (35% reduction in average) in *Histoplasma capsulatum*. In the ascomycetes *M. oryzae*, the model gene eGFP as well as two *M. oryzae* endogenous genes, polyketide synthase-like gene (PKS) and xylanase gene (XYL), were successfully silenced using the pSilent-Dual vector (*Figure 9.2B*). Even though the efficiency of silencing induced by pSilent-dual-based vectors is generally lower than that by pSilent-1-based vectors, strong silencing (> 80% reduction) was induced in a certain portion (3–5%) of transformants (Quoc *et al.*, unpublished data). In addition, silencing of the target genes was induced in wild-type but not the dicer mutant, indicating that pSilent-dual induces gene silencing through the RNAi pathway. Similarly, in the basidomycetes *C. neoformans*, the endogenous marker genes *URA5* and *ADE2* as well as other genes, *LAC1*, *CAP59* were effectively silenced using the pIBP48 vector, which has two opposing Gal7 promoters (I. Bose and T.L. Doering, personal communication).

9.2.3 Direct delivery of dsRNA into fungal cells

A method for direct delivery of dsRNA into protoplasts has recently been reported in *Phytophthora infestans,* which belongs to the fungus-like oomycetes (Stramenopiles) (Whisson *et al.*, 2005). A marker gene, GFP, and two *P. infestans* genes, *inf1* and *cdc14*, were transiently silenced by Lipofectin-mediated transfection of protoplasts with *in vitro* synthesized Cy3-labeled dsRNA (150–300 bp in size). GFP fluorescence was clearly reduced in regenerating protoplasts up to 4 days after exposure to GFP dsRNA. From 4 days after exposure, GFP fluorescence partially recovered but remained at a significantly reduced level (7–67% of the controls) until 17 days after transfection. When the stage-specific gene *cdc14* was silenced by transfection with homologous dsRNA, the expected phenotype of reduced numbers of sporangia was observed. In contrast, when the highly expressed *P. infestans* gene *inf1* was targeted for RNAi using this method, a significant reduction (11–44% of the controls) in *inf1* mRNA expression was detected only at 15 days but not 10 or 20 days after transfection. Interestingly, as assessed by mRNA abundance, the greatest level of silencing is observed from 12 to 15 days after transfection in most cases with *P. infestans*. In mammal systems, RNAi is typically activated within hours or 1–2 days, and remains effective for several days after transfection. Therefore, the apparent late occurrence of gene silencing, which is thought to involve signal amplification since the original trigger molecule should not be intact at this point, may be a characteristic of RNAi in oomycetes and possibly other fungi too.

9.2.4 Simultaneous silencing of multiple genes

In RNAi experiments, screening of silenced transformants or cell lines is a time-consuming process especially when silencing of the target gene produces no obvious phenotype. The technique of multiple gene silencing coupled with the use of a marker gene as an indication of the level of silencing may help solve this problem. Liu et al. (2002) first demonstrated in fungi that tandem silencing of two *C. neoformans* endogenous genes, *ADE2* and *CAP59* occurred by transformation with a single chimeric inverted repeat hairpin construct. More than 80% of the *ADE2*-silenced transformants also exhibited *CAP59*-silenced phenotype, indicating that this approach could provide an effective indicator to select transformants in which interference was operating. Similarly, in the apple scab fungus, *Venturia inaequalis*, two maker genes, GFP and endogenous trihydroxynaphthalene reductase (THN), which is involved in melanin biosynthesis, have been shown to be simultaneously silenced by a single chimeric construct (Fitzgerald *et al.*, 2004). Simultaneous silencing of both genes occurred at a frequency of 51% of all the transformants. The co-silencing frequency was slightly lower but comparable to that achieved by a hairpin construct for either the GFP (71%) or THN (61%) gene alone. Notably, the correlation between GFP and THN silencing was 100% when induced by the chimeric hairpin construct even though the level of silencing appeared to vary to some extent among the silenced transformants and between the two genes. This chimeric technology will be useful not only for screening of silenced transformants using an indicator gene such as GFP but also for co-silencing of multiple target genes by single transformation.

9.3 Genetic transformation and RNAi protocols for fungi

To induce RNAi in fungi, it is important to know how to introduce trigger molecules into living cells. Several transformation systems have been developed for a variety of fungal species, including the calcium chloride/polyethylene glycol (PEG) method for *Neurospora* (Case *et al.*, 1979), and *Aspergillus* (Yelton *et al.*, 1984) species, the lithium acetate method for *Saccharomyces* (Clancy *et al.*, 1984) and *Coprinus* species (Binninger *et al.*, 1987), electroporation for *Neurospora* and *Penicillium* species (Chakraborty *et al.*, 1991), microparticle bombardment for *Saccharomyces*, *Neurospora* (Armaleo *et al.*, 1990) and *Trichoderma* species (Lorito *et al.*, 1993), and more recently, *Agrobacterium tumefaciens*-mediated transformation (ATMT) for *Aspergillus* and other fungal species (de Groot *et al.*, 1998). Some of the abovementioned methods require protoplast preparation. Protoplast-based transformation such as the PEG method are well established and commonly used in model fungi such as *Aspergillus*, *Neurospora*, and *Magnaporthe*. However, protoplast preparation can be difficult in some fungal species because lytic enzymes do not always digest the cell wall sufficiently even if the mycelia are treated with a complex mixture of enzymes. In this section, four different protocols for inducing RNAi in fungi are described according to the original reports. They include the PEG method for *Magnaporthe oryzae* (Nakayashiki *et al.*, 2005), electroporation for *Cryptococcus neoformans* (Liu *et al.*, 2002), microparticle bombardment for *Mortierella alpina* (Takeno *et al.*, 2005), and Lipofectin-mediated transfection for *Phytophthora infestans* (Whisson *et al.*, 2005).

Acknowledgments

I am grateful to Tamara L. Doering, Washington University School of Medicine and Sakayu Shimizu and Eiji Sakuradani, Kyoto University, for critical review of the manuscript.

References

Aramayo, R. and Metzenberg, R.L. (1996) Meiotic transvection in fungi. *Cell* 6: 103–113.

Aravind, L., Watanabe, H., Lipman, D.J. and Koonin, E.V. (2000) Lineage-specific loss and divergence of functionally linked genes in eukaryotes. *Proc Natl Acad Sci USA* 97: 11319–11324.

Armaleo, D., Ye, G.N., Klein, T.M., Shark, K.B., Sanford, J.C. and Johnston, S.A. (1990) Biolistic nuclear transformation of *Saccharomyces cerevisiae* and other fungi. *Curr Genet* 17: 97–103.

Binninger, D.M., Skrzynia, C., Pukkila, P. J. and Casselton, L.A. (1987) DNA-mediated transformation of the basidiomycete *Coprinus cinereus*. *EMBO J* 6: 835–840.

Case, M.E., Schweizer, M., Kushner, S.R. and Giles, N.H. (1979) Efficient transformation of *Neurospora crassa* by utilizing hybrid plasmid DNA. *Proc Natl Acad Sci USA* 76: 5259–5263.

Catalanotto, C., Azzalin, G., Macino, G. and Cogoni, C. (2002) Involvement of small RNAs and role of the *qde* genes in the gene silencing pathway in *Neurospora*. *Genes Dev* 16: 790–795.

Catalanotto, C., Pallotta, M., ReFalo, P., Sachs, M.S., Vayssie, L., Macino, G. and Cogoni, C. (2004) Redundancy of the two dicer genes in transgene-induced posttranscriptional gene silencing in *Neurospora crassa*. *Mol Cell Biol* 24: 2536–2545.

Caten, C.E. and Jinks, J.L. (1968) Spontaneous variability of single isolates of *Phytophthora infestans*. I. Cultural variation. *Can J Bot* 46: 329–348.

Chakraborty, B.N., Patterson, N.A. and Kapoor, M. (1991) An electroporation-based system for high-efficiency transformation of germinated conidia of filamentous fungi. *Can J Microbiol* 37: 858–863.

Clancy, S., Mann, C., Davis, R.W. and Calos, M.P. (1984) Deletion of plasmid sequences during *Saccharomyces cerevisiae* transformation. *J Bacteriol* 159: 1065–1067.

Cogoni, C. and Macino, G. (1997) Isolation of quelling-defective (qde) mutants impaired in posttranscriptional transgene-induced gene silencing in *Neurospora crassa*. *Proc Natl Acad Sci USA* 94: 10233–10238.

Cogoni, C. and Macino, G. (1999) Gene silencing in *Neurospora crassa* requires a protein homologous to RNA-dependent RNA polymerase. *Nature* 399: 166–169.

Cogoni, C. and Macino, G. (2000) Post-transcriptional gene silencing across kingdoms. *Curr Opin Genet* 10: 638–643.

Cogoni, C., Irelan, J.T., Schumacher, M., Schmidhauser, T.J., Selker, E.U. and Macino, G. (1996) Transgene silencing of the *al-1* gene in vegetative cells of *Neurospora* is mediated by a cytoplasmic effector and does not depend on DNA-DNA interactions or DNA methylation. *EMBO J* 15: 3153–3163.

Dalmay, T., Hamilton, A., Rudd, S., Angell, S. and Baulcombe, D.C. (2000) An RNA-dependent RNA polymerase gene in *Arabidopsis* is required for posttranscriptional gene silencing mediated by a transgene but not by a virus. *Cell* 101: 543–553.

de Groot, M.J., Bundock, P., Hooykaas, P.J. and Beijersbergen, A.G. (1998) *Agrobacterium tumefaciens*-mediated transformation of filamentous fungi. *Nat Biotechnol* 16: 839–842.

de Jong, J.F., Deelstra, H.J., Wosten, H.A. and Lugones, L.G. (2006) RNA-mediated gene silencing in monokaryons and dikaryons of *Schizophyllum commune*. *Appl Environ Microbiol* **72**: 1267–1269.

Fagard, M., Boutet, S., Morel, J.B., Bellini, C. and Vaucheret, H. (2000) AGO1, QDE-2, and RDE-1 are related proteins required for post-transcriptional gene silencing in plants, quelling in fungi, and RNA interference in animals. *Proc Natl Acad Sci USA* **97**: 11650–11654.

Fitzgerald, A., van Kan, J.A.L. and Plummer, K.M. (2004) Simultaneous silencing of multiple genes in the apple scab fungus, *Venturia inaequalis*, by expression of RNA with chimeric inverted repeats. *Fungal Genet Biol* **41**: 963–971.

Galagan, J.E., Calvo, S.E., Borkovich, K.A., *et al.* (2003) The genome sequence of the filamentous fungus *Neurospora crassa*. *Nature* **422**: 859–868.

Giaever, G., Chu, A.M., Ni, L., *et al.* (2002) Functional profiling of the *Saccharomyces cerevisiae* genome. *Nature* **418**: 387–391.

Goldoni, M., Azzalin, G., Macino, G. and Cogoni, C. (2004) Efficient gene silencing by expression of double stranded RNA in *Neurospora crassa*. *Fungal Genet Biol* **41**: 1016–1024.

Hamada, W. and Spanu, P.D. (1988) Co-suppression of the hydrophobin gene HCf-1 is correlated with antisense RNA biosynthesis in *Cladosporium fulvum*. *Mol Gen Genet* **259**: 630–638.

Hammond, T.M. and Keller, N.P. (2005) RNA silencing in *Aspergillus nidulans* is independent of RNA-dependent RNA polymerases. *Genetics* **169**: 607–617.

Kadotani, N., Nakayashiki, H., Tosa, Y. and Mayama, S. (2003) RNA silencing in the phytopathogenic fungus *Magnaporthe oryzae*. *Mol Plant Microbe Interact* **16**: 769–776.

Latijnhouwers, M., Ligterink, W., Vleeshouwers, V.G.A.A., van West, P. and Govers, F. (2004) A Galpha subunit controls zoospore motility and virulence in the potato late blight pathogen *Phytophthora infestans*. *Mol Microbiol* **51**: 925–936.

Lee, D.W., Pratt, R.J., McLaughlin, M. and Aramayo, R. (2003) An argonaute-like protein is required for meiotic silencing. *Genetics* **164**: 821–828.

Liu, H., Cottrell, T.R., Pierini, L.M., Goldman, W.E. and Doering, T.L. (2002) RNA interference in the pathogenic fungus *Cryptococcus neoformans*. *Genetics* **160**: 463–470.

Lorito, M., Hayes, C. K., Di Pietro, A. and Harman, G.E. (1993) Biolistic transformation of *Trichoderma harzianum* and *Gliocladium virens* using plasmid and genomic DNA. *Curr Genet* **24**: 349–356.

Martens, H., Novotny, J., Oberstrass, J., Steck, T.L., Postlethwait, P. and Nellen, W. (2002) RNAi in *Dictyostelium*: the role of RNA-directed RNA polymerases and double-stranded RNase. *Mol Biol Cell* **13**: 445–453.

Martin-Hernandez, A.M., Dufresne, M., Hugouvieux, V., Melton, R. and Osbourn, A. (2000) Effects of targeted replacement of the tomatinase gene on the interaction of *Septoria lycopersici* with tomato plants. *Mol Plant–Microbe Interact* **13**: 1301–1311.

McDonald, T., Brown, D., Keller, N.P. and Hammond, T.M. (2005) RNA silencing of mycotoxin production in *Aspergillus* and *Fusarium* species. *Mol Plant Microbe Interact* **18**: 539–545.

Mourrain, P., Beclin, C., Elmayan, T., *et al.* (2000) *Arabidopsis* SGS2 and SGS3 genes are required for posttranscriptional gene silencing and natural virus resistance. *Cell* **101**: 533–542.

Mouyna, I., Henry, C., Doering, T.L. and Latge, J.P. (2004) Gene silencing with RNA interference in the human pathogenic fungus *Aspergillus fumigatus*. *FEMS Microbiol Lett* **237**: 317–324.

Nakayashiki, H. (2005) RNA silencing in fungi: mechanisms and applications. *FEBS Lett* **579**: 5950–5957.

Nakayashiki, H., Hanada, S., Nguyen, B.Q., Kadotani, N., Tosa, Y. and Mayama, S. (2005) RNA silencing as a tool for exploring gene function in ascomycete fungi. *Fungal Genet Biol* **42**: 275–283.

Nakayashiki, H., Kadotani, N. and Mayama, S. (2006) Evolution and diversification of RNA silencing proteins in fungi. *J Mol Evol* **63**: 127–135.

Namekawa, S.H., Iwabata, K., Sugawara, H., Hamada, F.N., Koshiyama, A., Chiku, H., Kamada, T. and Sakaguchi, K. (2005) Knockdown of LIM15/DMC1 in the mushroom *Coprinus cinereus* by double-stranded RNA-mediated gene silencing. *Microbiology* **151**: 3669–3678.

Nicolas, F.E., Torres-Martinez, S. and Ruiz-Vazquez, R.M. (2003) Two classes of small antisense RNAs in fungal RNA silencing triggered by non-integrative transgenes. *EMBO J* **22**: 3983–3991.

Rappleye, C.A., Engle, J.T. and Goldman, W.E. (2004) RNA interference in *Histoplasma capsulatum* demonstrates a role for alpha-(1,3)-glucan in virulence. *Mol Microbiol* **53**: 153–165.

Romano, N. and Macino, G. (1992) Quelling: transient inactivation of gene expression in *Neurospora crassa* by transformation with homologous sequences. *Mol Microbiol* **6**: 3343–3353.

Shiu, P.K.T., Raju, N.B., Zickler, D. and Metzenberg, R.L. (2001) Meiotic silencing by unpaired DNA. *Cell* **107**: 905–916.

Smardon, A., Spoerke, J.M., Stacey, S.C., Klein, M.E., Mackin, N. and Maine, E.M. (2000) EGO-1 is related to RNA-directed RNA polymerase and functions in germline development and RNA interference in *C. elegans*. *Curr Biol* **10**: 169–178.

Spiering, M.J., Moon, C.D., Wilkinson, H.H. and Schardl, C.L. (2005) Gene clusters for insecticidal loline alkaloids in the grass-endophytic fungus *Neotyphodium uncinatum*. *Genetics* **169**: 1403–1414.

Takeno, S., Sakuradani, E., Murata, S., Inohara-Ochiai, M., Kawashima, H., Ashikari, T. and Shimizu, S. (2004) Establishment of an overall transformation system for an oil-producing filamentous fungus, *Mortierella alpina* 1S-4. *Appl Microbiol Biotechnol* **65**: 419–425.

Takeno, S., Sakuradani, E., Tomi, A., Inohara-Ochiai, M., Kawashima, H., Ashikari, T. and Shimizu, S. (2005) Improvement of the fatty acid composition of an oil-producing filamentous fungus, *Mortierella alpina* 1S-4, through RNA interference with delta12-desaturase gene expression. *Appl Environ Microbiol* **71**: 5124–5128.

van West, P., Kamoun, S., van 't Klooster, J.W. and Govers, F. (1999) Internuclear gene silencing in *Phytophthora infestans*. *Mol Cell* **3**: 339–348.

Vollmer, S.J. and Yanofsky, C. (1986) Efficient cloning of genes of *Neurospora crassa*. *Proc Natl Acad Sci USA* **83**: 4869–4873.

Whisson, S.C., Avrova, A.O., van West, P. and Jones, J.T. (2005) A method for double-stranded RNA-mediated transient gene silencing in *Phytophthora infestans*. *Mol Plant Pathol* **6**: 153–163.

Wickes, B.L. and Edman, J.C. (1994) Development of a transformation system for *Cryptococcus neoformans*. In: *Molecular Biology of Pathogenic Fungi: A Laboratory Manual* (ed. B. Maresca and G.S. Kobayashi), pp. 309–313. Telos Press, New York.

Yelton, M.M., Hamer, J.E. and Timberlake, W.E. (1984) Transformation of *Aspergillus nidulans* by using a trpC plasmid. *Proc Natl Acad Sci USA* **81**: 1470–1474.

Protocol 9.1: Transformation of *Magnaporthe oryzae* by the calcium chloride/polyethylene glycol (PEG) method

MATERIALS

- CM liquid medium: 0.3% Casamino acids, 0.3% yeast extract, 0.5% sucrose. Autoclave 20 min at 121°C
- CM regeneration agar medium: CM liquid medium with 0.7% agar and 20% sucrose. Autoclave 20 min at 121°C
- Selective agar medium: CM liquid medium with 1.0% agar. Autoclave 20 min at 121°C. Hygromycin B is added after autoclave at a concentration of 400 µg/ml
- Digestion buffer: 10 mM Na_2HPO_4, 1.2 M $MgSO_4$ with 10 mg/ml Lysing enzymes (Sigma). Filtrate on 0.22 µm filter
- STC buffer: 1 M sorbitol, 50 mM Tris–HCl, pH 8.0, 50 mM $CaCl_2$. Autoclave 20 min at 121°C
- PEG solution: 60% PEG 3350 (Sigma) in 50 mM Tris–HCl, pH 8.0, 50 mM $CaCl_2$. Autoclave 20 min at 121°C

EQUIPMENT

- 50°C water bath
- Microcentrifuge
- Table-top centrifuge
- 26°C Rotary shaker
- 26°C incubator

PROCEDURE

1. PEG-mediated transformation of *M. oryzae* is performed as originally described by Vollmer and Yanofsky (1986) with some modifications. This protocol is applicable to model fungi such as *Neurospora*, *Aspergillus*, and *Fusarium* with appropriate modifications

2. Inoculate 50 ml of CM liquid medium with mycelia from an agar culture of *M. oryzae*, and incubate on a rotary shaker (200 rpm) at 26°C for 4–5 days

3. Harvest fungal mycelia by filtering through a Buchner funnel containing filter paper (Whatman no.1)

4. Add mycelial pad to 10 ml of digestion buffer per gram wet weight of mycelia in a sterile plastic tube, and incubate for 3 h at room temperature with gentle inversion on a shaker; tube in horizontal position

5. Filter digested protoplasts by gravity through four layers of sterile gauze, and centrifuge the filtrate in a swinging bucket rotor at 800 × g for 5 min

6. Pour off supernatant, and resuspend the collected protoplasts in 50 ml STC buffer

7. Centrifuge in a swinging bucket rotor at 800 × g for 5 min, pour off supernatant leaving just a little STC in the bottom, and resuspend in STC to obtain 1–5 × 10^7 protoplasts/ml

8. Add 100 μl of diluted protoplasts and 1–10 μg of plasmid DNA (ex. pSilent-1) to be transformed to 2-ml tubes

9. Mix gently and incubate on ice for 15 min

10. Add 200 μl, 400 μl and 600 μl of PEG solution in a step by step manner, and mix by gently pipetting up and down at each step

11. After incubation on ice for 10 min, collect protoplasts by centrifugation at 2000 × g for 3 min

12. Resuspend the protoplasts in 1 ml of STC, transfer in a 15-ml sterile plastic tube, add 10 ml of CM regeneration agar medium, and spread onto a sterilized petri dish

13. Incubate at 26°C overnight for the expression of the Hygromycine B resistant gene

14. The following day, add 5–10 ml of selective agar medium containing Hygromycin B (400 μg/ml) and incubate at 26°C until transformants appear

Protocol 9.2: Transformation of *Cryptococcus neoformans* by electroporation

MATERIALS

- YEPD: 1% yeast extract, 2% Bacto Peptone, 2% dextrose. Autoclave 20 min at 121°C
- EB Buffer (ice-cold): 10 mM Tris–HCl, pH 7.5, 1 mM $MgCl_2$, 270 mM sucrose in ultrapure water. Filtrate on 0.22 \µm filter
- Ice-cold sterile water

EQUIPMENT

- Rotary shaker at 30°C
- Table-top centrifuge
- BioRad electroporation device (settings 1 kV, 400 W, 25 µF)
- 0.2-cm BioRad electroporation cuvettes
- Microcentrifuge

PROCEDURE

1. Transformation of *C. neoformans* is performed as described by Wickes and Edman (1994) with some modifications

2. Inoculate 30 ml of YEPD medium with 5×10^5 cells of *C. neoformans*, and incubate overnight at 30°C on a rotary shaker

3. Dilute to 2×10^6 cells/ml or less using 100–200 ml of fresh YEPD broth and incubate at 30°C with vigorous shaking for 5–6 h

4. Harvest at around 6×10^6 to 1×10^7 cells/ml and collect cells by centrifugation for 5 min at $3000 \times g$ at 4°C

5. Wash cells with chilled water twice by centrifugation as above

6. Resuspend cells in 50 ml of chilled EB and add 200 µl of 1 M dithiothreitol (DTT)

7. Incubate on ice for 5–15 min, and collect cells by centrifugation as above

8. Wash cells in 50 ml EB without DTT and discard supernatant leaving ~1 ml of EB in the bottom

9. Resuspend cells in the EB remaining and dispense cell suspension into sterile 1.5 ml microfuge tubes, such that each tube contains ~3 × 10^8 cells

10. Centrifuge for 1 min at top-speed of a microfuge at 4°C, and remove supernatant, leaving ~ 0.04 ml of EB to resuspend cells

11. Add linearized DNA (1–5 µg) to the cell suspension, and transfer into pre-chilled cuvettes on ice

12. Electroporate with BioRad instrument using the following settings: 0.5 kV, 25 µF, and either 1000 or α resistance. The pulse length or time constants of 15–25 ms give a reasonable number of transformations

13. Add 1 ml of appropriate medium (ice-cold) immediately to the cuvette and transfer to a pre-cooled sterile 1.5 ml microfuge tube

14. Plate directly on appropriate medium. If selection is auxotrophic, use synthetic medium lacking specific nutrients. For a drug resistance marker, incubate the cells in rich medium for 1–2 h at 30°C, then plate on medium containing the selection drug

15. Incubate at 30°C until transformants appear

Protocol 9.3: Transformation of *Mortierella alpina* by the microparticle bombardment method

MATERIALS

- Czapek–Dox agar medium: 3% sucrose, 0.2% $NaNO_3$, 0.1% K_2HPO_4, 0.05% KCl, 0.05% $MgSO_4 \cdot 7H_2O$, 0.001% $FeSO_4 \cdot 7H_2O$, and 1.5% agar per liter. Adjust pH to 6.0. Autoclave 20 min at 121°C

- Uracil-free SC agar medium: 1.7 g Difco Yeast Nitrogen Base (w/o amino acids and ammonium sulfate), 5 g $(NH_4)_2SO_4$, 20 g glucose, 20 mg of adenine, 30 mg of tyrosine, 1 mg of methionine, 2 mg of arginine, 2 mg of histidine, 4 mg of lysine, 4 mg of tryptophan, 5 mg of threonine, 6 mg of isoleucine, 6 mg of leucine, 6 mg of phenylalanine and 20 g agar per liter. Autoclave 20 min at 121°C

EQUIPMENT

- Biolistic PDS-1000/He particle delivery system (Bio-Rad)

- Hepta adapter, biolistic

- Biolistic macrocarriers

- 1.1-µM tungsten beads

- Rupture discs

- Hepta stop screen

- 28°C incubator

- Microcentrifuge

PROCEDURE

Strain preparation

1. Transformation of *M. alpina* is performed as described by Takeno *et al.* (2004)

2. Culture a uracil auxotrophic *M. alpina* strain on Czapek–Dox agar medium containing 0.05 mg/ml uracil

3. Harvest intact spores of *M. alpina* from the surface of the medium (1.5×10^9 spores/300 cm^2)

4. Spread 1.5×10^8 spores on a petri dish containing uracil-free SC agar medium

DNA preparation

1. Prepare 30 mg of tungsten beads (1.1 μm in diameter) in a 1.5-ml tube as follows: wash once with 1 ml of 70% ethanol, wash twice with 1 ml of sterilized water, and add 500 μl of 50% glycerol (final concentration: 60 mg/ml tungsten beads)

2. The beads can be stored up to 2 weeks at 4°C

3. Aliquot beads into sterile tubes after vortexing for 5 min, and add the following in order while vortexing: 5 μl of DNA (1 mg/ml); 50 μl of 2.5 M $CaCl_2$; 20μl of 0.1 M spermidine

4. Continue vortexing for 2–3 min, and incubate at room temperature for 5 min

5. Spin for 5 s in a microfuge, and remove supernatant using a pipetman to avoid disturbing the pellet

6. Wash the pellet once with 140 μl of 70% ethanol and once with 140 μl of 100% ethanol

7. Resuspend the pellet in 48 μl of 100% EtOH

8. Vortex for at least 3 min, continue to resuspend by racking and vortexing until loaded on the macrocarriers

Bombardment

1. Place a petri dish with *M. alpina* spores in the holder on bottom slot in the PDS-1000/He Particle Delivery System (Bio-Rad)

2. Transfer 6 μl of beads onto a macrocarrier and spread it around with your pipette tip

3. Allow EtOH to evaporate to completion

4. Shoot twice at 1100 psi (7580 kPa)

5. Incubate at 28°C for 2–5 days, and transfer putative transformants to fresh SC agar medium

Protocol 9.4: Transformation of *Phytophthora infestans* by the Lipofectin-mediated transfection method

MATERIALS

- Rye A Agar medium (Caten and Jinks, 1968):
  - Soak 60 g of rye grains in 100 ml or less distilled water for 24 h at room temperature
  - Cover tray tightly with aluminum foil
  - Pour off and reserve supernatant
  - Blend the swollen grains for 2 min (distilled water may be added), and incubate for 1 h at 68°C in water bath
  - Filter through four thickness of gauze and discard the sediment
  - Combine the original supernatant with the filtrate, and add 20 g sucrose, 15 g Bacto Agar, then adjust volume to 1 l
  - Autoclave 20 min at 121°C
- Pea broth cultures: 125 g frozen fresh peas/l, boiled for 1 h, and filtered. Autoclave for 20 min at 121°C
- Digestion buffer: 1 M mannitol, 7 mM $MgSO_4$ with 5 mg/ml Lysing enzymes (Sigma) and 2.5 mg/ml cellulase (Sigma). Filtrate on 0.22-μm filter
- Osmoticum: 1 M mannitol, 7 mM $MgSO_4$
- Sterile distilled water

EQUIPMENT

- Rotary shaker
- Thermal cycler
- Microcentrifuge
- Table-top centrifuge
- 16°C incubator
- 20°C incubator
- 37°C incubator

PROCEDURE

Preparing protoplasts

1. Culture *P. infestans* on rye-A agar with rifampicin (30 μg/ml) and pimaricin (10 μg/ml) at 20°C for 2–3 weeks

2. Harvest sporangia from the agar culture by rubbing mycelium with a glass rod in 15 ml sterile distilled water

3. Inoculate 200 ml of pea broth cultures with dislodged sporangia, and incubate in the dark for 48 h at 20°C

4. Harvest mycelia by filtration through Miracloth (Calbiochem)

5. Add mycelia to 10 ml of digestion buffer per gram wet weight of mycelia, and incubate in a sterile plastic tube for 45 min at room temperature with gentle shaking (60 rpm); tube in horizontal position

6. Filter digested protoplasts by gravity through sterile Miracloth, and centrifuge the filtrate in a swinging bucket rotor at $600 \times g$ for 5 min

7. Pour off supernatant, and resuspend the collected protoplasts in 50 ml osmoticum

8. Wash four times with osmoticum by centrifugation as above

9. Resuspend the protoplasts in osmoticum at 1×10^5 protoplasts/ml

Preparing dsRNA and transfection

1. Amplify a 150–300-bp fragment of a target gene by PCR using two pairs of primers; a forward primer paired with a reverse primer with a T7 promoter sequence (5′-GTAATACGACTCAC-TATAGGG) at the 5′ end, and the same forward primer with an added 5′-T7 promoter sequence paired with the same reverse primer (no T7)

2. Use 5 μg of PCR product each for in vitro transcription of sense and antisense RNA using the Megascript RNAi kit (Ambion)

3. Incubate the reaction mixture for 16 h at 37°C

4. Mix synthesized sense and antisense RNAs to yield dsRNA

5. Remove remaining single-stranded RNA and DNA template by nuclease digestion as described in the Megascript RNAi kit protocol

6. Recover dsRNA by ethanol precipitation followed by centrifugation at $16\,000 \times g$, and dry completely

7. Resuspend the dry dsRNA pellet in osmoticum to yield a concentration of 4 μg/ml

8. Mix 10µl of dsRNA solution with an equal volume of Lipofectin reagent (Invitrogen) and incubate for 15 min at 20°C

9. Add 20 µl of protoplast solution (2000 protoplasts), mix gently and incubate for 24 h at 20°C

10. Add the entire mixture to 200 ml of pea broth with ampicillin and vancomycin (50 µg/ml each)

11. Transfer 2 ml of the broth into each well of a 24-well culture tray and incubate for 4 days at 20°C

12. Transfer individual regenerated colonies to agar medium, and conduct phenotype analyses

Index

Page numbers in *italics* indicate figures or tables.